The
SCIENCE
of
DISCWORLD
II

THE GLOBE

D1541723

The SCIENCE of DISCWORLD II
THE GLOBE

TERRY PRATCHETT
IAN STEWART & JACK COHEN

EBURY
PRESS

First published in Great Britain in 2002

This edition published in 2003

13

Text © Terry and Lyn Pratchett, Joat Enterprises, Jack Cohen 2002

Terry Pratchett, Ian Stewart and Jack Cohen have asserted their right to
be identified as the author of this work under the Copyright, Designs and
Patents Act 1988.

First published by
Ebury Press
Random House, 20 Vauxhall Bridge Road, London SW1V 2SA

The Random House Group Limited Reg. No. 954009

www.randomhouse.co.uk

Addresses for companies within The Random House Group
can be found at: www.randomhouse.co.uk

A CIP catalogue record for this book is available from the British Library

Cover Design by Dan Newman at Perfect Bound
Illustrations by Paul Kidby

ISBN 9780091888053

The Random House Group Limited supports The Forest Stewardship Council
(FSC®), the leading international forest certification organisation. Our books
carrying the FSC label are printed on FSC® certified paper. FSC is the only
forest certification scheme endorsed by the leading environmental organisations,
including Greenpeace. Our paper procurement policy can be found at
www.randomhouse.co.uk/environment

Printed and bound in the UK by
CPI Cox & Wyman, Reading, RG1 8EX

You spotted snakes with double tongue,
Thorny hedgehogs, be not seen;
Newts and blind-worms, do no wrong,
Come not near our fairy Queen.

I have had a most rare vision. I have had a dream, past the
wit of man to say what dream it was. Man is but an ass if he
go about to expound this dream. Methought I was— there is
no man can tell what. Methought I was, and methought I had,
but man is but a patch'd fool, if he will offer to say what
methought I had. The eye of man hath not heard, the ear of
man hath not seen, man's hand is not able to taste, his tongue
to conceive, nor his heart to report, what my dream was.

This is the silliest stuff that ever I heard.

WILLIAM SHAKESPEARE
A MIDSUMMER NIGHT'S DREAM

I'm nae listenin' to them! They've got warts!

ARTHUR J. NIGHTINGALE
THE SHORT COMEDY OF MACBETH

APOLOGY: this book is a true account of events in the life of William
Shakespeare, but only for a given value of 'true'.

Warning: *May Contain Nuts*

CONTENTS

ONE

MESSAGE IN A BOTTLE

 IN THE AIRY, CROWDED SILENCE of the forest, magic was hunting magic on silent feet.

A wizard may be safely defined as a large ego which comes to a point at the top. That is why wizards do not blend well. That would mean looking like other people, and wizards do not wish to look like other people. Wizards *aren't* other people.

And therefore, in these thick woods, full of dappled shade, new growth and birdsong, the wizards who were in theory blending in, in fact blended *out*. They'd understood the theory of camouflage – at least they'd nodded when it was being explained – but had then got it wrong.

For example, take this tree. It was short, and it had big gnarly roots. There were interesting holes in it. The leaves were a brilliant green. Moss hung from its branches. One hairy loop of grey-green moss, in particular, looked rather like a beard. Which was odd, because a lump in the wood above it looked rather like a nose. And then there was a blemish in the wood that could have been eyes …

But overall this was definitely a tree. In fact, it was a lot more like a tree than a tree normally is. Practically no other tree in the forest looked so tree-like as this tree. It projected a sensation of extreme barkness, it exuded leafidity. Pigeons and squirrels were queuing up to settle in the branches. There was even an owl. Other trees were just sticks with greenery on compared to the sylvanic verdancy of this tree …

… which raised a branch, and shot another tree. A spinning orange ball spun through the air and went splat! on a small oak.

Something happened to the oak. Bits of twig and shadows and bark which had clearly made up an image of a gnarled old tree now equally clearly became the face of Archchancellor Mustrum Ridcully, Master of Unseen University (for the extremely magical) and running with orange paint.

'Gotcha!' shouted the Dean, causing the owl to leap from his hat. This was lucky for the owl, because a travelling glob of blue paint removed the hat a moment later.

'Ahah! Take that, Dean!' shouted an ancient beech tree behind him as, changing without actually changing, it became the figure of the Lecturer in Recent Runes.

The Dean spun around, and a blob of orange paint hit him in the chest.

'Eat permitted colourings!' yelled an excited wizard.

The Dean glared across the clearing to a crabapple tree which was, now, the Chair of Indefinite Studies.

'What? I'm on *your* side, you damn fool!' he said.

'You can't be! You made such a good target!'*

The Dean raised his staff. Instantly, half a dozen orange and blue blobs exploded all over him as other hidden wizards let loose.

Archchancellor Ridcully wiped paint out of his eyes.

'All right, you fellows,' he sighed. 'Enough's enough for today. Time for tea, eh?'

It was so hard, he reflected, to get wizards to understand the concept of 'team spirit'. It simply wasn't part of wizardly thinking. A wizard could grasp the idea of, say, wizards versus some other group, but they lost their grip when it came to the idea of wizards against wizards. *Wizard* against wizards, yes, they had *no* trouble with that.

They'd start out as two teams, but as soon as there was any engagement they'd get all excited and twitchy and shoot other wizards indiscriminately. If you were a wizard then, deep down, you knew

---

* And in this short statement may be seen the very essence of wizardry.

that every other wizard was your enemy. If their wands had been left unfettered, rather than having been locked to produce only paint spells – Ridcully had been very careful about that – then this forest would have been on fire by now.

Still, the fresh air was doing them good. The University was far too stuffy, Ridcully had always thought. Out here there was sun, and bird-song, and a nice warm breeze—

—a cold breeze. The temperature was plunging.

Ridcully looked down at his staff. Ice crystals were forming on it.

'Turned a bit nippy all of a sudden, hasn't it?' he said, his breath tingling in the frigid air.

And then the world changed.

Rincewind, Egregious Professor of Cruel and Unusual Geography, was cataloguing his rock collection. This was, these days, the ground state of his being. When he had nothing else to do, he sorted rocks. His predecessors in the post had spent many years bringing back small examples of cruel or unusual geography and had never had time to catalogue them, so he saw this as his duty. Besides, it was wonderfully dull. He felt that there was not enough dullness in the world.

Rincewind was the least senior member of the faculty. Indeed, the Archchancellor had made it clear that in seniority terms he ranked somewhat lower than the things that went 'click' in the woodwork. He got no salary and had complete insecurity of tenure. On the other hand, he got his laundry done free, a place at mealtimes and a bucket of coal a day. He also had his own office, no one ever visited him and he was strictly forbidden from attempting to teach anything to anyone. In academic terms, therefore, he considered himself pretty lucky.

An additional reason for this was that he was in fact getting *seven* buckets of coal a day and so much clean laundry that even his socks were starched. This was because no one else had realised that Blunk, the coal porter, who was far too surly to read, delivered the buckets strictly according to the titles on the study doors.

The Dean, therefore, got one bucket. So did the Bursar.

Rincewind got seven because the Archchancellor had found him a useful recipient of all the titles, chairs and posts which (because of ancient bequests, covenants and, in one case at least, a curse) the University was obliged to keep filled. In most instances no one knew what the hell they were for or wanted anything to do with them, in case some clause somewhere involved students, so they were given to Rincewind.

Every morning, therefore, Blunk stoically delivered seven buckets to the joint door of the Professor of Cruel and Unusual Geography, the Chair of Experimental Serendipity, the Reader in Slood Dynamics, the Fretwork Teacher,* the Chair for the Public Misunderstanding of Magic, the Professor of Virtual Anthropology and the Lecturer in Approximate Accuracy ... who usually opened the door in his under-pants – that is to say, opened the door in the wall whilst wearing his underpants – and took the coal happily, even if it was a sweltering day. At Unseen University you had budgets, and if you didn't use up everything you'd been given you wouldn't get as much next time. If this meant you roasted all summer in order to be moderately warm during the winter, then that was a small price to pay for proper fiscal procedures.

On this day, Rincewind carried the buckets inside and tipped the coal on the heap in the corner.

Something behind him went 'gloink'.

It was a small, subtle and yet curiously intrusive sound, and it accompanied the appearance, on a shelf above Rincewind's desk, of a beer bottle *where no beer bottle had hitherto been.*

He took it down and stared at it. It had recently contained a pint of Winkle's Old Peculiar. There was absolutely nothing ethereal about it, except that it was blue. The label was the wrong colour and full of spelling mistakes but it was mostly there, right down to the warning in tiny, tiny print: May Contain Nuts.†

---

* This one was apparently the result of a curse some 1,200 years ago by a dying Archchancellor, which sounded very much like 'May you always teach fretwork!'

† Lord Vetinari, the Patrician and supreme ruler of the city, took proper food labelling

Now it contained a note.

He removed this with some care, and unrolled it, and read it.

Then he stared at the thing *beside* the beer bottle. It was a glass globe, about a foot across, and contained, floating within it, a smaller blue-and-fluffy-white globe.

The smaller globe was a world, and the space inside the globe was infinitely large. The world and indeed the whole universe of which it was part had been created by the wizards of Unseen University more or less by accident, and the fact that it had ended up on a shelf in Rincewind's tiny study was an accurate indication of how interested they were in it once the initial excitement had worn off.

Rincewind watched the world, sometimes, through an omniscope. It mostly had ice ages, and was less engrossing than an ant farm. Sometimes he shook it up to see if it would make it interesting, but this never seemed to have much of an effect.

Now he looked back at the note.

It was extremely puzzling. And the university had someone to deal with things like that.

Ponder Stibbons, like Rincewind, also had a number of jobs. However, instead of aspiring to seven, he perspired at three. He had long been the Reader in Invisible Writings, had drifted into the new post as Head of Inadvisably Applied Magic and had walked in all innocence into the office of Praelector, which is a university title meaning 'person who gets given the nuisance jobs'.

That meant that he was in charge in the absence of the senior members of the faculty. And, currently, this being the spring break,

---

very seriously. Unfortunately, he sought the advice of the wizards of Unseen University on this one, and posed the question thusly: 'Can you, taking into account multi-dimensional phase space, meta-statistical anomaly and the laws of probability, guarantee that anything *with absolute certainty* contains no nuts at all?' After several days, they had to conclude that the answer was 'no'. Lord Vetinari refused to accept 'Probably does not contain nuts' because he considered it unhelpful.

they *were* absent. And so were the students. The University was, therefore, running at near peak efficiency.

Ponder smoothed out the beer-smelling paper and read:

> TELL STIBBONS GET HERE AT ONCE. BRING LIBRARIAN. WAS IN FOREST, AM IN ROUNDWORLD. FOOD GOOD, BEER AWFUL. WIZARDS USELESS. ELVES HERE TOO. DIRTY DEEDS AFOOT.
>
> RIDCULLY

He looked up at the humming, clicking, *busy* bulk of Hex, the University's magical thinking engine, then, with great care, he placed the message on a tray that was part of the machine's rambling structure.

A mechanical eyeball about a foot across lowered itself carefully from the ceiling. Ponder didn't know how it worked, except that it contained vast amounts of incredibly finely drawn tubing. Hex had drawn up the plans one night and Ponder had taken them along to the gnome jewellers; he'd long ago lost track of what Hex was doing. The machine changed almost on a daily basis.

The write-out began to clatter and produced the message:

+++ Elves have entered Roundworld. This is to be expected. +++

'Expected?' said Ponder.

+++ Their world is a parasite universe. It needs a host +++

Ponder turned to Rincewind. 'Do you understand any of this?' he said.

'No,' said Rincewind. 'But I've run into elves.'

'And?'

'And then I've run away from them. You don't hang around elves. They're not my field, unless they're doing fretwork. Anyway, there's nothing *on* Roundworld at the moment.'

'I thought you did a report on the various species that kept turning up there?'

'You *read* that?'

'I read all the papers that get circulated,' said Ponder.

'You *do*?'

14

'You said that every so often some kind of intelligent life turns up, hangs around for a few million years, and then dies out because the air freezes or the continents explode or a giant rock smacks into the sea.'

'That's right,' said Rincewind. 'Currently the globe is a snowball again.'

'So what is the faculty doing there now?'

'Drinking beer, apparently.'

'When the whole world is frozen?'

'Perhaps it's lager.'

'But they are supposed to be running around in the woods, pulling together, solving problems and firing paint spells at one another,' said Ponder.

'What for?'

'Didn't you read the memo he sent out?'

Rincewind shuddered. 'Oh, I never read *those*,' he said.

'He took everyone off into the woods to build a dynamic team ethos,' said Ponder. 'It's one of the Archchancellor's Big Ideas. He says that if the faculty gets to know one another better, they'll be a happier, more efficient team.'

'But they *do* know one another! They've known one another for ages! That's why they don't like one another very much! They won't stand for being turned into a happy and efficient team!'

'Especially on a ball of ice,' said Ponder. 'They're supposed to be in woods fifty miles away, not in a glass globe in your study! There is no *way* to get into Roundworld without using a *considerable* amount of magic, and the Archchancellor has banned me from running the thaumic reactor at anything like full power.'

Rincewind looked again at the message from the bottle.

'How did the bottle get out?' he said.

Hex printed:

+++ I did that. I still maintain a watch on Roundworld. And I have been developing interesting procedures. It is now quite easy for me to reproduce an artefact in the real world +++

'Why didn't you tell us the Archchancellor needed help?' sighed Ponder.

+++ They were having such fun trying to send the bottle +++

'Can't you just bring them out, then?'

+++ Yes +++

'In that case—'

'Hold on,' said Rincewind, remembering the blue beerbottle and the spelling mistakes. 'Can you bring them out *alive?*'

Hex seemed affronted.

+++ Certainly. With a probability of 94.37 per cent +++

'Not great odds,' said Ponder, 'But perhaps—'

'Hold on again,' said Rincewind, still thinking about that bottle. 'Humans aren't bottles. How about alive, with fully functioning brains and all organs and limbs in the right place?'

Unusually, Hex paused before replying.

+++ There will be unavoidable minor changes +++

'How minor, exactly.'

+++ I cannot guarantee reacquiring more than one of every organ+++

There was a long, chilly silence from the wizards.

+++ Is this a problem? +++

'Maybe there's another way?' said Rincewind.

'What makes you think that?'

'The note asks for the Librarian.'

In the heat of the night, magic moved on silent feet.

One horizon was red with the setting sun. This world went around a central star. The elves did not know this and, if they had done, it would not have bothered them. They never bothered with detail of that kind. The universe had given rise to life in many strange places, but the elves were not interested in that, either.

This world had created lots of life. Up until now, none of it had ever had what the elves considered to be potential. But this time, there was definite promise.

Of course, it had iron, too. The elves hated iron. But this time, the rewards were worth the risk. This time …

One of them signalled. The prey was close at hand. And now they saw it, clustered in the trees around a clearing, dark blobs against the sunset.

The elves assembled. And then, at a pitch so strange that it entered the brain without the need to use the ears, they began to sing.

TWO

THE UMPTY-UMPTH ELEMENT

 DISCWORLD RUNS ON MAGIC, Roundworld runs on rules, and even though magic needs rules and some people think rules are magical, they are quite different things. At least, in the absence of wizardly interference. This was the main scientific message of our last book, *The Science of Discworld*. There we charted the history of the universe from the Big Bang through to the creation of the Earth and the evolution of a not especially promising species of ape. The story ended with a final fast-forward to the collapse of the space elevator by which a mysterious race (which could not *possibly* have been those apes, who were only interested in sex and mucking about) had escaped from the planet. They had left the Earth because a planet is altogether too dangerous a place to live, and had headed out into the galaxy in search of safety and a long-term chance of a decent pint.

The Discworld wizards never found out who the builders of the space elevator on Roundworld were. *We* know that they were *us*, the descendants of those apes, who'd brought sex and mucking about to high levels of sophistication. The wizards missed that bit, although, to be fair, the Earth had been in existence for over four billion years, and apes and humans were present only for a tiny percentage of that time. If the entire history of the universe were compressed to one day, we would have been present for the final 20 seconds.

Quite a lot of interesting things happened on Roundworld while the wizards were skipping ahead, and now, in this present book, the wizards are going to find out what those things were. And of course they're going to interfere, and inadvertently create the world we live in today, just as their interference in the Roundworld Project inadvertently created our entire universe. It has to work like that, doesn't it?

That's how the story goes.

Seen from outside, as it sits in Rincewind's office, the entire human universe is a small sphere. Large quantities of magic went into its manufacture and, paradoxically, into maintaining its most interesting feature. Which is this: Roundworld is the only place on Discworld where magic does not work. A strong magical field protects it from the thaumic energies that surge around it. Inside Roundworld, things don't happen because people want them to or because they make a good narrative: they happen because the rules of the universe, the so-called 'laws of nature', *make* them happen.

At least, that was a reasonable way to describe things … until human beings evolved. At that point, something very strange happened to Roundworld. It began, in various ways, to resemble Discworld. The apes acquired minds, and their minds started to interfere with the normal running of the universe. Things started to happen *because* human minds wanted them to. Suddenly the laws of nature, which up to that point had been blind, mindless rules, were infused with purpose and intention. Things started to happen for a reason, and among these things that happened was reasoning itself. Yet this dramatic change took place without the slightest violation of the same rules that had, up to that point, made the universe a place without purpose. Which, on the level of the rules, it still is.

This seems like a paradox. The main content of our scientific commentary, interleaved between successive episodes of a Discworld story, will be to resolve that paradox: how did Mind (capital 'M' for 'metaphysical') come into being on this planet? How did a Mindless universe 'make up its own Mind'? How can we reconcile human free will (or its semblance) with the inevitability of natural law? What is the relation between the 'inner world' of the mind and the allegedly objective 'outer world' of physical reality?

The philosopher René Descartes argued that the mind must be built from some special kind of material – 'mind-stuff' that was different from ordinary matter, indeed undetectable using ordinary matter. Mind was an invisible spiritual essence that animated otherwise unthinking matter. It was a nice idea, because it explained at a stroke why Mind is so strange, and for a long time it was the conventional view. Nevertheless, today this concept of 'Cartesian duality' has fallen out of favour. Nowadays only cosmologists and particle physicists are allowed to invent new kinds of matter when they want to explain why their theories totally fail to match observed reality. When cosmologists find that galaxies are rotating at the wrong speeds in the wrong places, they don't throw away their theories of gravitation. They invent 'cold dark matter' to fill in the missing 90 per cent of the mass of the universe. If any other scientists did that kind of thing, people would throw up their hands in horror and condemn it as 'theory saving'. But cosmologists seem to get away with it.

One reason is that this idea has many advantages. Cold dark matter is cold, dark and material. Cold means that you can't detect it by the heat radiation that it throws off, because it doesn't. Dark means that you can't detect it by the light that it emits, because it doesn't. Matter means that it's a perfectly ordinary material *thing* (not some silly invention like Descartes' immaterial mind-stuff). Having said that, of course, cold dark matter is totally invisible, and it's definitely not the same as conventional matter, which isn't cold and isn't dark ...

To their credit, the cosmologists are trying very hard to find a way to detect cold dark matter. So far, they've discovered that it does bend light, so you can 'see' lumps of cold dark matter by the effect they have on images of more distant galaxies. Cold dark matter creates mirage-like distortions in the light from distant galaxies, smearing them out into thin arcs, centred on the lump of missing mass. From those distortions, astronomers can re-create the distribution of that otherwise invisible cold dark matter. The first results are coming in now, and within a few years it will be possible to survey the universe and find out whether the missing 90 per cent of matter really is there, cold and dark as expected, or whether the whole idea is nonsense.

Descartes' similarly invisible, undetectable mind-stuff has had a very different history. At first, its existence seemed obvious: minds simply do not behave like the rest of the material world. Then, its existence seemed obvious nonsense, because you can chop a brain into pieces, preferably after ensuring that its owner has previously departed this world, and look for its material constituents. And when you do, there's nothing unusual there. There's lots of complicated proteins, arranged in very elaborate ways, but you won't find a single atom of mind-stuff.*

We can't yet dissect a galaxy, so for now cosmologists can get away with their absurd invention of a face-saving new material. Neuroscientists, trying to explain the mind, have no such luxury. Brains are much easier to pull apart than galaxies.

Despite the change in current conventional wisdom, there remain a few diehard dualists who still believe in special mind-stuff. But today, nearly all neuroscientists believe that the secret of Mind lies in the structure of the brain, and even more importantly, in the processes that the brain carries out. As you read these words, you experience a strong sense of Self. There is a You that is doing the reading, and thinking about the words and the ideas they express. No scientist has ever dissected out the bit of the brain that contains this impression of You. Most suspect that no such bit exists: instead, you feel like You because of the overall activity of your entire brain, plus the nerve fibres that are connected to it, bringing it sensations of the outside world and allowing it to control the movement of your arms, legs and fingers. You feel like you, in fact, because you are busily *being* You.

Mind is a process carried out within a brain made of perfectly ordinary matter, in accordance with the rules of physics. It is, however, a very strange process. There is a kind of duality, but it is a duality of interpretation rather than of physical material. When you think a thought – about, let us say, the Fifth Elephant that slipped off the back of Great A'Tuin, orbited in an arc of a circle and crashed on to the surface of the Discworld – the same physical act of thinking that thought has two distinct meanings.

One of them is straightforward physics. In your brain, various electrons are surging to and fro in various nerve fibres. Chemical molecules

* And you'd be in the position of the horrible Discworld 'Auditors', who are anthropomorphic representations of the rules of the universe, who in *Thief of Time* reduce paintings and statues to their component atoms in a fruitless search for 'beauty'.

are combining together, or breaking up, to make new ones. Modern sensing apparatus, such as the PET scanner,* can reconstruct a three-dimensional image of your brain, showing which regions are active when you are thinking about that elephant. Materially, your brain is buzzing in some complicated way. Science can see how it is buzzing, but it can't (yet) extract the elephant.

That's the second interpretation. From inside, so to speak, you have no sensation of those buzzing electrons and reacting chemicals. Instead, you have a very vivid impression of a large grey creature with flappy ears and a trunk, sailing improbably through space and crashing disastrously to the ground. Mind is what it feels like to *be* a brain. The same physical events acquire a totally different meaning when viewed from the inside. One task of science is to try to bridge the gap between those two interpretations. The first step is to figure out which bits of the brain do what when you think a particular thought. To reconstruct, in fact, the elephant from the electrons. That's not yet possible, but every day brings it a step closer. Even when science gets there, it will probably not be able to explain why your impression of that elephant is so vivid, or why it takes exactly the form that it does.

In the study of consciousness there is a technical term for what a perception 'feels like'. It is called a *quale* (pronounced 'kwah-lay', not 'quail'), a figment that our minds paint on to their model of the universe in the way that an artist adds pigment to a portrait. Such qualia (plural) paint the world in vivid colours so that we can respond more quickly to it, and, in particular, respond to signs of danger, food, possible sexual partners ... Science has no explanation of why qualia feel like they do, and it's not likely to get one. So science can explain how a mind works, but not what it is like to be one. No shame in that: after all, physicists can explain how an electron works, but not what it is like to be one. Some questions are beyond science. And, we suspect, beyond anything else: it is easy enough to *claim* an explanation of these metaphysical problems, but just as impossible to prove you're

* PET = Positron Emission Tomography, meaning that the machine picks up tiny particles emitted by the tissues of the brain and reconstructs a map of what's going on inside it.

right. Science admits it can't handle these things, so at least it's honest.

At any rate, the science of the mind (small 'M' now because we're not talking metaphysics) addresses how the mind works, and how it evolved, but not what it's like to be one. Even with this limitation, the science of the brain is not the whole story. There is another important dimension to the question of Mind. Not how the brain works and what it does, but how it came to be like that.

How, on Roundworld, did Mind evolve from mindless creatures?

Much of the answer lies not inside the brain, but in its interactions with the rest of the universe. Especially other brains. Human beings are social animals, and they communicate with each other. The trick of communication made a huge, qualitative change to the evolution of the brain and its ability to house a mind. It accelerated the evolutionary process, because the transfer of ideas happens much faster than the transfer of genes.

How do we communicate? We tell stories. And that, we shall argue, is the real secret of Mind. Which brings us back to Discworld, because on Discworld things really do work the way human minds *think* they do on Roundworld. Especially when it comes to stories.

Discworld runs on magic, and magic is indissolubly linked to Narrative Causality, the power of story. A spell is a story about what a person wants to happen, and magic is what turns stories into reality. On Discworld, things happen *because people expect them to*. The sun comes up every day because that's its job: it was set up to provide light for the people to see by, and it comes up during the day when people need it. That's what suns do; that's what they're *for*. And it's a proper, *sensible* sun, too: a smallish fire not very far away, which goes over and under the Disc, incidentally but entirely logically causing one of the elephants to lift a leg to let it pass. It's not the ridiculous, pathetic kind of sun that we have – absolutely gigantic, infernally hot, and nearly a hundred million miles away because it's too dangerous to be near. And we go round it instead of it going round us, which is crazy, especially since what every human being on the planet sees, other than the visually impaired, is the latter. It's a terrible waste of material just to make daylight ...

On Discworld, the eighth son of an eighth son *must* become a wizard. There's no escaping the power of story: the outcome is inevitable. Even if, as in *Equal Rites*, the eighth son of an eighth son is a girl. Great A'Tuin the turtle must swim though space with four elephants on its back and the entire Discworld on top of them, because that's what a world-bearing turtle has to do. The narrative structure demands it. Moreover, on Discworld everything that there is* exists as a *thing*. To use the philosophers' language, concepts are *reified*: made real. Death is not just a process of cessation and decay: he is also a person, a skeleton with a cloak and a scythe, and he TALKS LIKE THIS. On Discworld, the narrative imperative is reified into a substance, narrativium. Narrativium is an element, like sulphur or hydrogen or uranium. Its symbol ought to be something like Na, but thanks to a bunch of ancient Italians that's already reserved for sodium (so much for So). So it's probably Nv, or maybe Zq given what they've done to sodium. Be that as it may, narrativium is an element on Discworld, so it lives somewhere in the Disc's analogue of Dmitri Mendeleev's periodic table. Where? The Bursar of Unseen University, the only wizard insane enough to understand imaginary numbers, would doubtless tell us that there is no question: it is the umpty-umpth element.

Discworld narrativium is a substance. It takes care of narrative imperatives, and ensures they are obeyed. On Roundworld, our world, humans act as if narrativium exists here, too. We expect it not to rain tomorrow *because* the village fair is on, and it would be unfair (in both senses) if rain spoiled the occasion.

Or, more often, given the pessimistic ways of our country folk, we expect it to *rain* tomorrow because the village fair is on. Most people expect the universe to be mildly malevolent but hope it will be kindly disposed, whereas scientists expect it to be indifferent. Drought-struck farmers pray for rain, in the express hope that the universe or owner thereof will hear their words and suspend the laws of meteorology for their benefit. Some, of course, actually believe just that, and for all anyone can prove, they could be right. This is a tricky question, and a delicate one; let us just say that no reputable scientific observer has

---

* And many things that there aren't, such as Dark.

yet caught God breaking the laws of physics (although of course He might be too clever for them) and leave it at that for the moment.

And this is where Mind takes centre stage.

The curious thing about the human belief in narrativium is that once humans evolved on the planet, their beliefs started to be true. We have, in a way, created our own narrativium. It exists in our minds, and there it is a process, not a thing. On the level of the material universe, it's just one more pattern of buzzing electrons. But on the level of what it feels like to be a mind, it operates just like narrativium. Not only that: it operates on the material world, not just the mental one: its *effects* are just like those of narrativium. Generally our minds control our bodies – sometimes they don't, and indeed sometimes it's the other way round, especially during adolescence – and our bodies make things happen out there in the material world. Within each person there is a 'strange loop', which confuses the mental and material levels of existence.

This strange loop has a curious effect on causality. We get up in the morning and leave the house at 7.15 *because* we have to get to work by 9 o'clock. Scientifically, this is a very bizarre form of causality: the future is affecting the past. That doesn't normally occur in physics (except in very esoteric Quantum things, but let's not get distracted). In this case, science has an explanation. What causes you to get up at 7.15 is *not* actually your future arrival at work. If in fact you fall under a bus and never make it to work, you still got up at 7.15. Instead of backwards causality, you have a mental model, in your brain, which is your best attempt to predict the day ahead. In that model, realised as buzzing electrons, you *think* that you ought to be at work by nine. That model, and its expectation of the future, exists *now*, or more accurately, a short time in the past. It is that expectation that causes you to get up instead of lying in and having a well-deserved snooze. And the causality is entirely normal: from past to future by way of actions taking place in the present.

So that's all right then. Except that when you think of it, the causality is still very strange. A few electrons, buzzing in ways that are meaningless from the outside of the brain in which they reside, lead to a coherent action by a 70-kilogram lump of protein. Well, at that time in the morning it's not a *very* coherent lump of protein, but you understand what we mean. That's why we call this very creative piece of confusion a strange loop.

Those mental models are stories, simplified narratives that correspond in a rough-hewn way to aspects of the world that we consider to be important. Note that 'we': all mental models are infected with human biases. Our minds tell us stories about the world, and we base a great many of our actions on what those stories say. Here, the story is 'the person who arrived at work late and was fired from their job'. That story alone will lever us out of bed at an unearthly hour, even if we get on well with the boss and fondly imagine that the story doesn't apply to *us*. In other words, we make up our world according to the stories that we tell ourselves, and each other, about it.

We build minds in our children that way, too. The Western child is brought up on stories like the time Winnie the Pooh went to Rabbit's house, ate too much honey and got stuck in the entrance hole on the way out.* The story tells us not to be too greedy; that terrible things will happen to us if we are. Even the child knows that Winnie the Pooh is fiction, but they understand what the story is about. It doesn't lead them to avoid pigging out on honey, and it doesn't make them worry about getting stuck in the doorway when they try to leave the room after having eaten too much dinner. The story isn't about literal interpretations. It's a metaphor, and the mind is a metaphor machine.

The power of narrativium in Roundworld is immense. Things happen because of it that you would never expect from the laws of nature. For example, the laws of nature pretty much forbid an Earthbound object suddenly leaping up into space and landing on the Moon. They don't say it's impossible, but they do imply that you could wait a very long time indeed before it happened. Despite this, there is a machine on the Moon. Several. They all used to be down here. They are there because, centuries ago, people told each other romantic tales about the Moon. She was a goddess, who looked down on us. When full, she caused werewolves to change from humans into animals. Even

---

* It *would* have been an exit hole, but he didn't.

then, humans were quite good at doublethink; the Moon was clearly a big silver disc, but, *at the same time*, she was a goddess.

Slowly those tales changed. Now the Moon was another world, and by harnessing the power of swans we could fly there in a chariot. Then (Jules Verne suggested) we could get there in a hollowed-out cylinder fired by a giant gun, located in Florida. Finally, in the 1960s, we found the right kind of swan (liquid oxygen and hydrogen) and the right kind of chariot (several million tons of metal) and we flew to the Moon. In a hollowed-out cylinder, launched from Florida. It wasn't exactly a gun. Well, actually it was in a basic physical sense; the rocket was the gun and it went along for the ride, firing burnt fuel in place of a bullet.

If we'd not told ourselves stories about the Moon, there would have been no point in going there at all. An interesting view, maybe ... but we 'knew' about the view only because we had told ourselves scientific stories about images sent back by space probes. Why did we go? Because we'd been telling ourselves that we would, one day, for several hundred years. Because we'd made it inevitable and introduced it into the 'future story' of a great many people. Because it satisfied our curiosity, and because the Moon was waiting. The Moon was a story waiting to be finished ('First human lands on the Moon!'), and we went there because the story demanded it.

When Mind evolved on Earth, a kind of narrativium evolved alongside it. Unlike the Discworld variety of narrativium, which on the Disc is just as real as iron or copper or praseodymium, our variety is purely mental. It is an imperative, but the imperative has not been reified into a thing. However, we have the sort of mind, that respond to imperatives, and to many other non-things. And so it *feels* to us as if our universe runs on narrativium.

There is a curious resonance here, and 'resonance' is definitely the word. Physicists tell a story about how carbon forms in the universe. In certain stars there is a particular nuclear reaction, a 'resonance' between nearby energy levels, which gives nature a stepping-stone from lighter elements to carbon. Without that resonance, so the story goes, carbon could not have formed. Now, the laws of physics as we currently understand them involve several 'fundamental constants', such as the speed of light, Planck's constant in quantum theory, and the charge on an electron. These numbers determine the quantitative implications of the physical laws, but any choice of constants sets up a potential universe. The way that a universe behaves depends on the actual numbers that are used in its laws. As it happens, carbon is an essential constituent of all known life. All of which leads up to a clever little story called the Anthropic Principle: that it's silly for us to ask why we live in a universe whose physical constants make that nuclear resonance possible – because if we didn't, there'd be no carbon, hence no us to ask about it.

The story of the carbon resonance can be found in many science books, because it creates a powerful impression of hidden order in the universe, and it seems to explain so much. But if we look a little more closely at this story, we find that it is a beautiful illustration of the seductive power of a compelling but false narrative. When a story seems to hang together, even consciously self-critical scientists can fail to ask the question that makes it fall apart.

Here's how the story goes. Carbon is created in red giant stars by a rather delicate process of nuclear synthesis, called the triple-alpha process. This involves the fusion of three helium nuclei.* A helium nucleus contains two protons and two neutrons. If you fuse three helium nuclei together, you get six protons and six neutrons. That, as it happens, is a carbon nucleus.

All very well, but the odds on such a triple collision occurring inside a star are very small. Collisions of two helium nuclei are much more common, though still relatively rare. It is extremely rare for a third helium nucleus to crash into two that are just colliding. It's like paint-balls and wizards. Every so often, a paintball will go *splat!* against a wizard. But you wouldn't bet a lot of money on a second paintball hitting him at the exact same moment. This means that the synthesis of

---

* In the simplest picture of an atom, the nucleus is a relatively small central region made from protons and neutrons. Electrons 'orbit' the nucleus at a distance. The triple-alpha process takes place in a plasma, where the atoms have been stripped of their electrons, so only their nuclei are involved. Later, as the plasma cools, the nuclei can acquire the necessary electrons.

carbon has to take place in a series of steps rather than all at once, and the obvious way is for two helium nuclei to fuse, and then for a third helium nucleus to fuse with the result.

The first step is easy, and the resulting nucleus has four protons and four neutrons: this is one form of the element beryllium. However, the lifetime of this particular form of beryllium is only 10^{-16} seconds, which gives that third helium nucleus a very small target to aim at. The chance of hitting this target is incredibly small, and it turns out that the universe hasn't existed long enough for even a tiny fraction of its carbon to have been made in this way. So triple collisions are out, and carbon remains a puzzle.

Unless … there is a loophole in the argument. And indeed there is. The fusion of beryllium with helium, leading to carbon, would occur much more rapidly, yielding a lot more carbon in a much shorter time, if the energy of carbon just happens to be close to the combined energies of beryllium and helium. This kind of near-equality of energies is called a *resonance*. In the 1950s Fred Hoyle insisted that carbon has to come from somewhere, and predicted that there must therefore exist a resonant state of the carbon atom. It had to have a very specific energy, which he calculated must be about 7.6 MeV.*

Within a decade, it was discovered that there is a state with energy 7.6549 MeV. Unfortunately, it turns out that the combined energies of beryllium and helium are about 4 per cent higher than this. In nuclear physics, that's a huge error.

Oops.

Ah, but, miraculously, that apparent discrepancy is just what we want. Why? Because the additional energy imparted by the temperatures found in a red giant star is exactly what's needed to change the combined energy of beryllium and helium nuclei by that missing 4 per cent.

---

* 1 MeV is one million electron-volts. An electron-volt is a unit of energy, obviously, and for our current purposes it doesn't really matter what that unit *is*. For the record, it's the energy of an electron when its potential is raised by one volt, and is equal to 1.6×10^{-12} ergs. And the energy referred to here is the excess energy compared to the lowest energy state of the atom, its 'ground state'. What's an erg? Look it up if you really need to know.

Wow.

It's a wonderful story, and it rightly earned Hoyle huge numbers of scientific brownie-points. And it makes our existence look rather delicate. If the fundamental constants of the universe are changed, then so is that vital 7.6549. So it is tempting to conclude that our universe's constants are fine-tuned for carbon, making it very special indeed. And it is equally tempting to conclude that the reason for that fine-tuning is to ensure that complex life turns up. Hoyle didn't do that, but many other scientists have given into these temptations.

Sounds good: what's wrong? The physicist Victor Stenger calls this kind of argument 'cosmythology'. Another physicist, Craig Hogan, has put his finger on one of the weak points. The argument treats the temperature of the red giant and that 4 per cent discrepancy in energy levels as if they were independent. That is, it assumes that you can change the fundamental constants of physics without changing the way a red giant works. However, that's obvious nonsense. Hogan points out that 'the structure of stars includes a built-in thermostat that automatically adjusts the temperature to just the value needed to make the reaction go at the correct rate'. It's rather like being amazed that the temperature in a fire is just right to burn wood, when in fact that temperature is caused by the chemical reaction that burns the wood. This kind of failure to examine the interconnectedness of natural phenomena is a typical, and quite common, error in anthropic reasoning.

In the human world, what counts is not carbon, but narrativium. And in that context we wish to state a new kind of anthropic principle. It so happens that we live in a universe whose physical constants are just right for carbon-based brains to evolve to the point at which they create narrativium, much as a star creates carbon. And the narrativium does crazy things, like putting machines on the Moon. Indeed, if carbon did not (yet) exist, then any narrativium-based lifeform could find some way to manufacture it, by telling itself a really gripping story about the need for carbon. So causality in this universe is irredeemably weird. Physicists like to put it all down to the fundamental constants, but it's more likely an example of Murphy's law.

But that's another story.

The more we think about narrative in human affairs, the more we see that our world revolves around the power of story. We build our minds by telling stories. Newspapers select news according to its value as a story, not according to how intrinsically important it is. 'England loses cricket match to Australia' is a story (though not a very surprising one) and it goes on the front page. 'Doctors think that they may have improved the diagnosis of liver disease by 1 per cent' is not a story, even though most science works like that (and in years to come, depending on the state of your liver, you might think it's a rather more important story than a cricket match).

'Scientist claims cure for cancer' *is* a story, though, even if the supposed cure is nonsense. So are 'spiritualist medium claims a cure for cancer', and 'Secret code predictions hidden in the Bible', more's the pity.

As we write, there is a furore over a small group of people who are proposing to clone a human being. It's a major story, but very few newspapers are reporting the most likely result of this attempt, which will be abject failure. It took 277 failures, many rather nasty, before Dolly the Sheep was cloned, and she has now been found to have serious genetic defects, poor lamb.

Trying to clone a human may indeed be unethical, but that's not the best reason for objecting to this misguided and foolish attempt. The best reason is that it won't work, because nobody yet knows how to overcome numerous technical obstacles; moreover, if by some stroke of (mis)fortune it *did* happen to work, any child produced would have serious defects. Producing such a child, now that *is* unethical.

Making 'carbon copies' of human beings, which is the usual basis of the newspapers' *story* about the ethics, is beside the point. That's not what cloning does, anyway. Dolly the Sheep was *not* genetically identical to her mother, though she came close. Even if she had been, she would still have been a different sheep, moulded by different experiences. For that matter, the same would be true even if she *was* genetically identical to her mother. For the same reason, cloning a dead child will not bring *that* child back to life. Much of the media discussion of the ethics of cloning, like much of the public understanding of science, is vaguely stirred through with science fiction. In this arena, as in so many, the power of the story outweighs any questions about the real factual basis.

Human beings do not just tell stories, or just listen to them. They are more like Granny Weatherwax, who is aware of the power of story on Discworld, and refuses to be trapped by the story's narrativium. Instead, she *uses* the power of story to mould events according to her own wishes. Roundworld priests, politicians, scientists, teachers and journalists have learned to use the power of story to get their messages across to the public, and to manipulate or persuade people to behave in particular ways. The 'scientific method' is a defence mechanism against that kind of manipulation. It tells you not to believe things because you want them to be true. The proper scientific response to any new discovery or theory, especially your own, is to look for ways to disprove it. That is, to try to find a different story that explains the same things.

The anthropologists got it wrong when they named our species *Homo sapiens* ('wise man'). In any case it's an arrogant and bigheaded thing to say, wisdom being one of our least evident features. In reality, we are *Pan narrans*, the storytelling chimpanzee.

At this point, the structure of *The Science of Discworld 2: The Globe* becomes very self-referential. You will need to bear that in mind as we proceed. The book is itself a story – no – two intertwined stories. One, the odd-numbered chapters, is a Discworld fantasy. The other, the even-numbered chapters, is a story of the science of the Mind (metaphysical again). The two are closely related, designed to fit together like foot and glove;* the science story is presented as a series of Very Large Footnotes to the fantasy story.

So far, so good … but it gets more complicated. When you read a Discworld story, you play a curious mental game. You react as if the story is true, as if Discworld actually exists, as if Rincewind and the Luggage are real, and Roundworld is but a fragment of a long-forgotten dream. (Please stop interrupting, Rincewind, we know it's different from your point of view. Yes, of course we're the ones that

---

* Not hand and glove, the fit isn't *that* close.

don't exist, we're bundles of rules whose consequences take place only inside a small globe on a dusty shelf in Unseen University. Yes, we do appreciate that, and will you please *shut up*?) Sorry about that.

People have become very good at playing this game, and we will exploit that by setting Earth and Discworld on the same narrative level, so that each illuminates the other. In the first book, *The Science of Discworld*, the Discworld defined what is real. That's why reality makes such good sense. Roundworld is a magical construct, designed to keep the magic out, and that's why it makes no sense at all (to wizards, at least). In this sequel Earth acquires inhabitants, the inhabitants acquire minds, and minds do strange things. They bring narrativium to a story-less universe.

A computer can do a billion sums in the blink of a keystroke and get them all right, but it couldn't pretend to be a cowardly wizard if one walked up to it and thumped it on the memory cache. In contrast, we can think ourselves inside the mind of a cowardly wizard with ease, or recognise someone else when they're acting the part of one, but we're completely lost when it comes to doing several million simple sums a second. Even though, to someone not of this universe, that might appear to be a simpler task.

That's because we run on narrativium, and computers don't.

THREE

JOURNEY INTO L-SPACE

IT WAS THREE HOURS LATER, in the cool of Unseen University. Not much had changed in the High Energy Magic building, except that a screen had been set up to show the output of Ponder's iconograph projector.

'I don't see why you need it,' said Rincewind. 'There's only the two of us.'

'Ook,' agreed the Librarian. He was annoyed at having been woken from a doze in his library. It had been a very *gentle* awakening, since no one wakes up a 300lb orangutan roughly (twice, at least) but he was still annoyed.

'The Archchancellor says that we've got to be more organised about these things,' said Ponder. 'He says it's no use just shouting out "Hey, I've got a great idea!" These things have got to be *presented* properly. Are you ready?'

The very small imp that ran the projector raised a tiny thumb.

'Very well,' said Ponder. 'First slide. This is the Roundworld as it currently—'

'It's the wrong way up,' said Rincewind.

Ponder looked at the image.

'It's a *ball*,' he snarled. 'It's *floating* in *space*. How can it be the wrong way up?'

'That crinkly continent should be at the top.'

'Very well!' snapped Ponder. 'Imp, turn it around. Right? Satisfied?'

'It's the right way up but now it's the wrong way arou—' Rincewind began.

34

There was a thwack as Ponder's pointer stick smacked into the screen. 'This is the Roundworld!' he snapped. 'As it exists at present! A world covered in ice! But time on Roundworld is subordinate to time in the real world! All times in Roundworld are accessible to us, in the same way that all pages in a book, though consecutive, are accessible to us! I have ascertained that the Faculty *are* on Roundworld but not in what appears to be the present time! They are several hundred million years in the past! Which is, from our point of view, perfectly capable of *also* being the present! I don't know how they got there! It should not be physically possible! Hex has located them! We have to assume that they can't get back the way they came! However – next slide please!'

Click!

'It's the same one,' said Rincewind. 'But now it's sideways—'

'A globe has no sideways!' said Ponder. There was a tinkle of breaking glass from the direction of the projector, and some very small cursing.

'I just thought you wanted to do it properly,' murmured Rincewind. 'Anyway, this is going to be about L-space, isn't it? I *know* it is. You know it is.'

'Yes, but I don't say that yet! I've got another dozen slides to come!' gasped Ponder. 'And a flow chart!'

'But it is, isn't it,' said Rincewind wearily. 'I mean, they say they've found other wizards. That means libraries. That means you can get there through L-space.'

'I was going to say that's how *we* can get there,' said Ponder.

'Yes, I know,' said Rincewind. 'That's why I thought I'd take the opportunity of saying "you" at this early juncture.'

'How can there be wizards on Roundworld?' said Ponder. 'When we *know* magic doesn't work there?'

'Search me,' said Rincewind. 'Ridcully did say they're useless.'

'And why can't the faculty come back by themselves? They were able to send the bottle! That must have used magic, surely?'

'Why not just go and ask them?' said Rincewind.

'You mean by homing in on the distinctive biothaumic signature of a group of wizards?'

'Well, I was thinking of waiting until something dreadful happened and *you* going to have a look in the wreckage,' said Rincewind. 'But the other stuff would probably work.'

'The omniscope locates them in approximately the 40,002,730,907th century,' said Ponder, staring at the globe. 'I can't get an image. But if we can find a way to the nearest library—'

'Ook!' said the Librarian. And then he ooked some more. He ooked at length, with an occasional eek. Once he thumped his fist on the table. He didn't need to thump the table a second time. There wasn't, at that point, much in the way of table left to thump.

'He says only very senior librarians can use L-space,' said Rincewind, as the Librarian folded his arms. 'He was quite emphatic. He says it's not to be treated like some kind of magic funfair ride.'

'But it's an order from the Archchancellor!' said Ponder. 'There isn't any other way to get there!'

The Librarian looked a little uncertain at that. Rincewind knew why. It was hard to be an orangutan in Unseen University, and the only way the Librarian had been able to deal with it was by acknowledging Mustrum Ridcully as the alpha male, even though the Archchancellor seldom climbed up to a high place on the rooftops and called mournfully over the city at dawn. This meant that, unlike the other wizards, he found it very hard to shrug off an archchancelloric command. It was a direct, fang-revealing, chest-beating challenge.

Rincewind had an idea.

'If we put the globe in the Library,' he said to the ape, 'then that would mean that even though you are travelling in L-space you would not be taking Mr Stibbons anywhere outside the Library. I mean, the globe would be inside the library, so even though you'd wind up in the globe, you really wouldn't have travelled very far at all. A few feet, maybe. The globe's only infinite on the inside, after all.'

'Well, Rincewind, I am *impressed*,' said Ponder, while the Librarian looked perplexed. 'I'd always thought of you as rather stupid, but that was a remarkable piece of verbal reasoning. If we put the globe down right on the Librarian's desk, say, then the whole journey would take place *inside* the library, right?'

'Exactly,' said Rincewind, who was prepared to overlook 'rather stupid' in view of this unexpected praise.

'And it's perfectly safe in the library, after all …'

'Big thick walls. Very safe place,' Rincewind agreed.

'So, put like that, no harm will come to us,' said Ponder.

'There you go with the "us" again,' said Rincewind, backing away.

'We'll find them and bring them back!' said Ponder. 'How hard can it be?'

'It can be incredibly hard! There's *elves* there! You know elves! They are *dangerous!* Drop your guard for a moment and they can control your mind!'

'They chased me through some woods once,' said Ponder. 'They are very frightening. I remember writing that down in my diary.'

'You *wrote down in your diary* that you were scared?'

'Yes. Why not? Don't you?'

'I haven't got a big enough diary. But it makes no sense! There's nothing on the Roundworld that elves would be interested in! They like to have … slaves. And we've never seen anything evolve that's bright enough to be a slave.'

'You might have missed something,' said Ponder.

'No, *I* say *you*, *you* say we,' said Rincewind.

They both stared at the globe.

'Look, it's like having a pot plant,' said Ponder. 'If it has greenfly, you try to squash them.'

'*I* never do that,' said Rincewind. 'Greenfly may be small, but there's a lot of them …'

'It was a metaphor, Rincewind,' said Ponder, wearily.

'… I mean, supposing they decide to gang up?'

'Rincewind, you are the only other person here who knows anything at all about Roundworld. You *will* come with us or … or … I'll tell the Archchancellor about the seven buckets.'

'How do you know about the seven buckets?'

'*And* I'll explain to him how all of your jobs could easily be done by a simple set of instructions for Hex, too. It'd take me about, oh, thirty seconds. Let's see …

\# Rincewind

```
SUB WAIT
WAIT
RETURN
```

Or possibly

```
RUN RINCEWIND'
```

'You wouldn't do that!' said Rincewind. 'Would you?'

'Certainly. Now, are you coming? Oh, and bring the Luggage.'

Knowledge = power = energy = matter = mass, and on that simple equation rests the whole of L-space. It is via L-space that all books are connected (quoting the ones before them, and influencing the ones that come after). But there is no time in L-space. Nor is there, strictly speaking, any space. Nevertheless, L-space is infinitely large and connects all libraries, everywhere and everywhen. It's never further than the other side of the bookshelf, yet only the most senior and respected librarians know the way in.

From inside, L-space looked to Rincewind like a library designed by someone who did not have to worry about time, budget, strength of materials or physics. There are some laws, though, that are coded into the very nature of the universe, and one is: There Is Never Enough Shelf Space.*

He turned and looked back. They'd entered L-space by walking *through* what had looked like a solid wall of books. He *knew* it was a solid wall, he'd taken books off those shelves before now. You had to be a very senior Librarian indeed to know in what precise circumstances you could step straight through it.

He could still see the library through the gap, but it faded from view as he watched. What remained was books. Mountains of books. Hills and valleys of books. Perilous precipices of books. Even in what

---

* Others found by research wizards include Objects In The Rear View Mirror Are Closer Than They Appear, No User Serviceable Parts Inside and, of course, May Contain Nuts.

passed for the sky, which was a sort of blue grey, there was a distant suggestion of books. There is never enough shelf space, anywhere.

Ponder was carrying a considerable amount of magical equipment. Rincewind, being a more experienced traveller, was carrying as little weight as possible. Everything else was being carried by the Luggage, which looked like a sea chest but with a number of pink, human-like and fully operational feet.

'Under the rules of the Roundworld, magic can't work,' said Ponder, as they followed the Librarian. 'Won't the Luggage stop existing?'

'It's worth a try,' said Rincewind, who felt that owning a semi-sapient and occasionally homicidal box on legs reduced his opportunities to make live friends, 'but it doesn't usually worry about rules. They bend round it. Anyway, it's already been there before, for a very long time, without any damage. To the Luggage, anyway.'

The walls of books shifted as the wizards approached; in fact, each step radically changed the nature of the bookscape which was in any case, said Ponder, a mere metaphorical depiction created by their brains to allow them to deal with the unimaginable reality. The shifting perspective would have given most people a serious headache at least, but Unseen University had rooms where the gravity moved around during the course of a day, one corridor of infinite length and several windows that only existed on one side of their walls. Life at UU reduced your capacity for surprise by quite a lot.

Occasionally the Librarian would stop, and sniff at the books nearest to him. At last he said 'ook', quietly, and pointed to another stack of books. There were, drawn gently on the spine of an old leather-bound volume, some chalk marks.

'Librarian-sign,' said Rincewind. 'He's been here before. We're close to Roundworld book-space.'

'How could he—' Ponder began, and then said: 'Oh. I see. Er ... Roundworld exists in L-space even before we created it? I mean *yes*, obviously I know that's true, but even so—'

Rincewind took a book from a pile near him. The cover was brightly coloured and made of paper, suggesting an absence of cows on the originating world, and had the title: *Sleep Well My Lovely Falcon*. The words inside made even less sense.

'It might not have been worth our trouble,' he said.

The Librarian said 'ook', which Rincewind understood as 'I'm going to get into real trouble with the Secret Masters of the Library for this day's work'.

Then the ape appeared to triangulate on the bookscape around him, and knuckled forward, and vanished.

Ponder looked at Rincewind. 'Did you see how he did that?' he said, and then a hairy red arm appeared out of the air and jerked him off his feet. A moment later the same thing happened to Rincewind.

It wasn't *much* of a library, but Rincewind knew how this worked. Two books were a library – for a lot of people, two books were an *enormous* library. But even one book could be a library, if it was a book that made a big enough dimple in L-space. A book with a title like *100 Ways with Broccoli* was unlikely to be one such, whereas *The Relationship Between Capital and Labour* might be, especially if it had an appendix on making explosives. The deeply magical and interminably ancient volumes in the Library of UU strained the fabric of L-space like a baby elephant on a worn-out trampoline, leaving it so thin that the Library was a potent and easy portal.

Sometimes, though, even one book could do that. Even one line. Even one word, in the right place and the right time.

The room was large, panelled and sparsely furnished. Quite a lot of paperwork was strewn on a desk. Quill pens lay by an inkwell. A window looked out on to broad gardens, where it was raining. A skull lent a homely touch.

Rincewind leaned down and tapped it.

'Hello?' he said. He looked up at the others.

'Well, the one in the Dean's office can sing comic songs,' he said defensively. He stared at the paperwork on the desk. It was covered in symbols which had a magical look, although he didn't recognise any of them. On the other side of the room, the Librarian was leafing through one of the books. Strangely, they weren't on shelves. Some were neatly piled, others locked in boxes, or at least in boxes that were locked until the Librarian tried to lift the lid.

Occasionally he pursed his lips and blew a disdainful raspberry.

'Ook,' he muttered.

'Alchemy?' said Rincewind. 'Oh dear. That stuff never works.' He lifted up what looked like a small leather hatbox, and removed the lid. 'This is more like it!' he said, and pulled out a ball of smoky quartz. 'Our man is definitely a wizard!'

'This is very bad,' said Ponder, staring at a device in his hand. 'Very, very bad indeed.'

'What is?' said Rincewind, turning around quickly.

'I'm reading a very high glamour quotient,' said Ponder.

'There's elves here?'

'Here? The place is practically elvish!' said Ponder. 'The Archchancellor was right.'

All three explorers stood quietly. The Librarian's nostrils flared. Rincewind sniffed, very cautiously.

'Seems okay to me,' he said, at last.

Then a man in black entered the room. He came in quickly, opening the door no more than necessary, in a kind of aggressive sidle, and stopped in astonishment. Then his hand flew to his belt and he drew a thin, businesslike sword.

He saw the Librarian. He stopped. And then it was really all over, because the Librarian could unfold his arm very fast and, importantly, there was a fist like a sledgehammer on the end of it.

As the dark figure slid down the wall, the crystal sphere in Rincewind's hand said: 'I believe I now have enough information. I advise departure from this place at a convenient opportunity and in any case before this gentleman awakes.'

'Hex?' said Ponder.

'Yes. Let me repeat my advice. Lack of absence from this place will undoubtedly result in metal entering the body.'

'But you're talking via a crystal ball! Magic doesn't work here!'

'Don't argue with a voice saying "run away"!' said Rincewind. 'That's good advice! You don't question it! Let's get out of here!'

He looked at the Librarian, who was sniffing along the bookshelves with a puzzled expression.

Rincewind had a sense for the universe's tendency to go wrong. He didn't leap to conclusions, he plunged headlong towards them.

'You've brought us out through a one-way door, haven't you ...' he said.

'Oook!'

'Well, how long will it take to find the way in?'

The Librarian shrugged and returned his attention to the shelves.

'Leave now,' said the crystal Hex. 'Return later. The owner of this house will be useful. But leave before Sir Francis Walsingham wakes up, because otherwise he will kill you. Steal his purse from him first. You will need money. For one thing, you will need to pay someone to give the Librarian a shave.'

'*Oook?*'

FOUR

THE ADJACENT POSSIBLE

 THE CONCEPT OF L-SPACE, short for 'Library-space', occurs in several of the Discworld novels. An early example occurs in *Lords and Ladies*, a story that is mostly about elvish evil. We are told that Ponder Stibbons is Reader in Invisible Writings, and this phrase deserves (and gets) an explanation:

> The study of invisible writings was a new discipline made available by the discovery of the bi-directional nature of Library-space. The thaumic mathematics are complex, but boil down to the fact that all books, everywhere, affect all other books. This is obvious: books inspire other books written in the future, and cite books written in the past. But the General Theory* of L-space suggests that, in that case, the contents of books *as yet unwritten* can be deduced from books now in existence.

L-space is a typical example of the Discworld habit of taking a metaphorical concept and making it real. The concept here is known as 'phase space', and it was introduced by the French mathematician Henri Poincaré about a hundred years ago to open up the possibility of applying geometrical reasoning to dynamics. Poincaré's metaphor has now invaded the whole of science, if not beyond, and we will

---

* There's a Special Theory as well, but no one bothers with it much because it's self-evidently a load of marsh-gas. [This footnote is a footnote in the original quotation. So this is a metafootnote.]

make good use of it in our discussion of the role of narrativium in the evolution of the mind.

Poincaré was the archetypal absent-minded academic – no, come to think of it he was 'present-minded somewhere else', namely in his mathematics, and it's easy to understand why. He was probably the most naturally gifted mathematician of the nineteenth century. If you had a mind like his, you'd spend most of *your* time somewhere else, too, revelling in the beauty of the mathiverse.

Poincaré ranged over almost all of mathematics, and he wrote several best-selling popular science books, too. In one piece of research, which single-handedly created a new 'qualitative' way of thinking about dynamics, he pointed out that when you are studying some physical system that can exist in a variety of different states, then it may be a good idea to consider the states that it *could* be in, but isn't, as well as the particular state in which it *is*. By doing that, you set up a context that lets you understand what the system is doing, and why. This context is the 'phase space' of the system. Each possible state can be thought of as a *point* in that phase space. As time passes, the state changes, so this representative point traces out a curve, the *trajectory* of the system. The rule that determines the successive steps in the trajectory is the *dynamic* of the system. In most areas of physics, the dynamic is completely determined, once and for all, but we can extend this terminology to cases where the rule involves possible choices. A good example is a game. Now the phase space is the space of possible positions, the dynamic is the rules of the game and a trajectory is a legal sequence of moves by the players.

The formal setting and terminology for phase spaces is not as important, for us, as the viewpoint that they encourage. For example, you might wonder why the surface of a pool of water, in the absence of wind or other disturbances, is flat. It just sits there, flat; it isn't even *doing* anything. But you start to make progress immediately if you ask the question 'what would happen if it *wasn't* flat?' For instance, why can't the water be piled up into a hump in the middle of the pond? Well, imagine that it was. Imagine that you can control the position of every molecule of water, and that you pile it up in this way, miraculously keeping every molecule just where you've placed it. Then, you 'let go'. What would happen? The heap of water would collapse, and waves would slosh across the pool until everything settled down to that nice, flat surface that we've learned to expect. Again, suppose you arranged the water so that there was a big dip in the middle. Then as soon as you let go, water would move in from the sides to fill the dip.

Mathematically, this idea can be formalised in terms of the space of all possible shapes for the water's surface. 'Possible' here doesn't mean *physically* possible: the only shape you'll ever see in the real world, barring disturbances, is a flat surface. 'Possible' means 'conceptually possible'. So we can set up this space of all possible shapes for the surface as a simple mathematical construct, and this is the phase space for the problem. Each 'point' – location – in phase space represents a conceivable shape for the surface. Just one of those points, one state, represents 'flat'.

Having defined the appropriate phase space, the next step is to understand the dynamic: the way that the natural flow of water under gravity affects the possible surfaces of the pool. In this case, there is a simple principle that solves the whole problem: the idea that water flows so as to make its total energy as small as possible. If you put the water into some particular state, like that piled-up hump, and then let go, the surface will follow the 'energy gradient' downhill, until it finds the lowest possible energy. Then (after some sloshing around which slowly subsides because of friction) it will remain at rest in this lowest-energy state.

The energy in this problem is 'potential energy', determined by gravity. The potential energy of a mass of water is equal to its height above some arbitrary reference level, multiplied by the mass concerned. Suppose that the water is not flat. Then some parts are higher up than others. So we can transfer some water from the high level to the lower one, by flattening a hump and filling a dip. When we do that, the water involved moves downwards, so the total energy decreases. Conclusion: if the surface is not flat, then the energy is not as small as possible. Or, to put it the other way round: the minimum energy configuration occurs when the surface is flat.

The shape of a soap bubble is another example. Why is it round? The way to answer that question is to compare the actual round shape with a hypothetical non-round shape. What's different? Yes, the alternative isn't round, but is there some less obvious difference? According to Greek legend, Dido was offered as much land (in northern Africa) as she could enclose with a bull's hide. She cut it into a very long, thin strip and enclosed a circle. There she founded the city of Carthage. Why did she choose a circle? Because the circle is the shape with greatest area, for a given perimeter. In the same way, a sphere is the shape with greatest volume, for a given surface area; or, to put it another way, it is the shape with the smallest surface area that contains a given volume. A soap bubble contains a fixed volume of air, and its surface area gives the energy of the soap film due to surface tension. In the space of all possible shapes for bubbles, the one with the least energy is a sphere. All other shapes have larger energy, and are therefore ruled out.

You may not feel that bubbles are important. But the same principle explains why Roundworld (the planet not the universe, but maybe that, too) is round. When it was molten rock, it settled into a spherical shape, because that had the least energy. For the same reason, the heavy materials like iron sank into the core, and the lighter ones, like continents and air, floated up to the top. Actually, Roundworld isn't exactly a sphere, because it rotates, so centrifugal forces cause it to bulge at the equator. But the amount of bulge is only one-third of one per cent. And that bulging shape is the minimum-energy configuration for a mass of liquid spinning at the same speed as the Earth's rotation when it was just starting to solidify.

The physics here isn't important for the message of this book. What is important is the 'Worlds of If' point of view involved in the application of phase spaces. When we discussed the shape of water in a pond, we pretty much ignored the flat surface, the thing we were trying to explain. The entire argument hinged upon non-flat surfaces, humps and dips, and hypothetical transfers of water from one to the other. Almost all of the explanation involved thinking about things that don't actually happen. Only at the end, having ruled out *all* non-flat surfaces, did we observe that the only possibility left was therefore what the water would actually do. The same goes for the bubble.

At first sight, this might seem to be a very oblique way of doing physics. It takes the stance that the way to understand the real world is to ignore it, and focus instead on all the possible alternative unreal worlds. Then we find some principle (in this case, minimum energy) to rule out nearly all of the unreal worlds, and see what's left. Wouldn't it be easier to start with the real world, and focus solely on that? No, it wouldn't. As we've just seen, the real world alone is too limited to offer a convincing explanation. What you get from the real world alone is 'the world is like it is, and there's nothing more to be said'. However, if you take the imaginative leap of considering unreal worlds, too, you can compare the real world with all of those unreal worlds, and maybe find a principle that picks out the real one from all the others. Then you have answered the question '*Why* is the world the way it is, rather than something else?'

An excellent way to approach 'why' questions is to consider alternatives and rule them out. 'Why did you park the car round the corner down a side-street?' 'Because if I'd parked outside the front door on the double yellow lines, a traffic warden would have given me a parking ticket.' This particular 'why' question is a story, a piece of fiction: a hypothetical discussion of the likely consequences of an action that never occurred. Humans invented their own brand of narrativium as an aid to the exploration of I-space, the space of 'insteads'. Narrative provides I-space with a geography: if I did this *instead* of that, then what would happen would be …

On Discworld, phase spaces are real. The fictitious alternatives to the one actual state exist, too, and you can get inside the phase space and roam over its landscape – provided you know the right spells, secret entrances and other magical paraphernalia. L-space is a case in point. On Roundworld, we can *pretend* that phase space exists, and we can imagine exploring its geography. This pretence has turned out to be extraordinarily insightful.

Associated with any physical system, then, is a phase space, a space of the possible. If you're studying the solar system, then the phase space comprises all possible ways to arrange one star, nine planets, a considerable number of moons and a gigantic number of asteroids in space. If you're studying a sand-pile, then the phase space comprises the number of possible ways to arrange several million grains of sand. If you're studying thermodynamics, then the phase space comprises all possible positions and velocities for a large number of gas molecules. Indeed, for each molecule there are three position coordinates and three velocity coordinates, because the molecule lives in three-dimensional space. So with N molecules there are $6N$ coordinates altogether. If you're looking at games of chess, then the phase space consists of all possible positions of the pieces on the board. If you're thinking about all possible books, then the phase space is L-space. And if you're thinking about all possible universes, you're contemplating U-space. Each 'point' of U-space is an entire universe (and you have to invent the multiverse to hold them all ...)

When cosmologists think about varying the natural constants, as we described in Chapter 2 in connection with the carbon resonance in stars, they are thinking about one tiny and rather obvious piece of U-space, the part that can be derived from our universe by changing the fundamental constants but otherwise keeping the laws the same. There are infinitely many other ways to set up an alternative universe: they range from having 101 dimensions and totally different laws to being identical with our universe except for six atoms of dysprosium in the core of the star Procyon that change into iodine on Thursdays.

As this example suggests, the first thing to appreciate about phase spaces is that they are generally rather big. What the universe actually does is a tiny proportion of all the things it could have done instead. For instance, suppose that a car park has one hundred parking slots, and that cars are either red, blue, green, white, or black. When the car park is full, how many different patterns of colour are there? Ignore the make of car, ignore how well or badly it is parked; focus solely on the pattern of colours.

Mathematicians call this kind of question 'combinatorics', and they have devised all sorts of clever ways to find answers. Roughly speaking, combinatorics is the art of counting things without actually counting them. Many years ago a mathematical acquaintance of ours came across a university administrator counting light bulbs in the roof of a lecture hall. The lights were arranged in a perfect rectangular grid, 10 by 20. The administrator was staring at the ceiling, going '49, 50, 51 ...'

'Two hundred,' said the mathematician.

'How do you know that?'

'Well, it's a 10 by 20 grid, and 10 times 20 is 200.'

'No, no,' replied the administrator. 'I want the *exact* number.'*

Back to those cars. There are five colours, and each slot can be filled by just one of them. So there are five ways to fill the first slot, five ways to fill the second, and so on. Any way to fill the first slot can be combined with any way to fill the second, so those two slots can be filled in $5 \times 5 = 25$ ways. Each of those can be combined with any of the five ways to fill the third slot, so now we have $25 \times 5 = 125$ possibilities. By the same reasoning, the total number of ways to fill the whole car park is $5 \times 5 \times 5 \ldots \times 5$, with a hundred fives. This is 5^{100}, which is rather big. To be precise, it is

$$78886090522101180541172856528278622$$
$$9673206435109023004770278930664625$$

(we've broken the number in two so that it fits the page width) which has 70 digits. It took a computer algebra system about five seconds to work that out, by the way, and about 4.999 of those seconds were taken up with giving it the instructions. And most of the rest was used up printing the result to the screen. Anyway, you now see why combinatorics is the art of counting without actually *counting*; if you listed all the possibilities and counted them '1, 2, 3, 4 ...' you'd never finish. So it's a good job that the university administrator wasn't in charge of car parking.

How big is L-space? The Librarian said it is infinite, which is true if you used infinity to mean 'a much larger number than I can envisage' or if you don't place an upper limit on how big a book can be,† or if

---

* The bean-counters don't even know how to count beans sensibly. Are we surprised?

† A tour of any airport bookshop will show that this is reasonable.

you allow all possible alphabets, syllabaries, and pictograms. If we stick to 'ordinary-sized' English books, we can reduce the estimate.

A typical book is 100,000 words long, or about 600,000 characters (letters and spaces, we'll ignore punctuation marks). There are 26 letters in the English alphabet, plus a space, making 27 characters that can go into each of the 600,000 possible positions. The counting principle that we used to solve the car-parking problem now implies that the maximum number of books of this length is $27^{600,000}$, which is roughly $10^{860,000}$ (that is, an 860,000-digit number). Of course, most of those 'books' make very little sense, because we've not yet insisted that the letters make sensible words. If we assume that the words are drawn from a list of 10,000 standard ones, and calculate the number of ways to arrange 100,000 words in order, then the figure changes to $10,000^{100,000}$, equal to $10^{400,000}$, and this is quite a bit smaller ... but still enormous. Mind you, most of those books wouldn't make much sense either; they'd read something like 'Cabbage patronymic forgotten prohibit hostile quintessence' continuing at book length.* So maybe we ought to work with sentences ... At any rate, even if we cut the numbers down in that manner, it turns out that the universe is not big enough to contain that many physical books. So it's a good job that L-space is available, and now we know why there's never enough shelf space. We like to think that our major libraries, such as the British Library or the Library of Congress, are pretty big. But, in fact, the space of those books that actually exist is a tiny, tiny fraction of L-space, of all the books that could have existed. In particular, we're never going to run out of new books to write.

Poincaré's phase space viewpoint has proved to be so useful that nowadays you'll find it in every area of science – and in areas that aren't science at all. A major consumer of phase spaces is economics. Suppose that a national economy involves a million different goods: cheese, bicycles, rats-on-a-stick, and so on. Associated with each good is a price, say £2.35 for a lump of cheese, £449.99 for a bicycle, £15.00 for a rat-on-a-stick. So the state of the economy is a list of one million

---

* But Joycean scholars would be furious if we excluded *Finnegan's Wake*, which reads *exactly* like that.

numbers. The phase space consists of all possible lists of a million numbers, including many lists that make no economic sense at all, such as lists that include the £0.02 bicycle or the £999,999,999.95 rat. The economist's job is to discover the principles that select, from the space of all possible lists of numbers, the actual list that is observed.

The classic principle of this kind is the Law of Supply and Demand, which says that if goods are in short supply and you really, really want them, then the price goes up. It sometimes works, but it often doesn't. Finding such laws is something of a black art, and the results are not totally convincing, but that just tells us that economics is hard. Poor results notwithstanding, the economist's way of thinking is a phase space point of view.

Here's a little tale that shows just how far removed economic theory is from reality. The basis of conventional economics is the idea of a rational agent with perfect information, who maximises utility. According to these assumptions, a taxi-driver, for example, will arrange his activities to generate the most money for the least effort.

Now, the income of a taxi-driver depends on circumstances. On good days, with lots of passengers around, he will do well; on bad days, he won't. A rational taxi-driver will therefore work longer on good days and give up early on bad ones. However, a study of taxi-drivers in New York carried out by Colin Camerer and others shows the exact opposite. The taxi-drivers seem to set themselves a daily target, and stop working once they reach it. So they work shorter hours on good days, and longer hours on bad ones. They could increase their earnings by 8 per cent just by working the same number of hours every day, for the same total working time. If they worked longer on good days and shorter on bad ones, they could increase their earnings by 15 per cent. But they don't have a good enough intuition for economic phase space to appreciate this. They are adopting a common human trait of placing too much value on what they have today, and too little on what they may gain tomorrow.

Biology, too, has been invaded by phase spaces. The first of these to gain widespread currency was DNA-space. Associated with every living organism is its genome, a string of chemical molecules called DNA. The DNA molecule is a double helix, two spirals wrapped round a common core. Each spiral is made up of a string of 'bases' or 'nucleotides', which come in four varieties: cytosine, guanine, adenine, thymine, normally abbreviated to their initials C, G, A, T. The sequences on the two strings are 'complementary': wherever C appears on one string, you get G on the other, and similarly for A and T. So the DNA contains two copies of the sequence, one positive and one negative, so to speak. In the abstract, then, the genome can be thought of as a single sequence of these four letters, something like AATG-GCCTCAG ... going on for rather a long time. The human genome, for example, goes on for about three billion letters.

The phase space for genomes, DNA-space, consists of all possible sequences of a given length. If we're thinking about human beings, the relevant DNA-space comprises all possible sequences of three billion code letters C, G, A, T. How big is that space? It's the same problem as the cars in the car park, mathematically speaking, so the answer is $4 \times 4 \times 4 \times ... \times 4$ with three billion 4s. That is, $4^{3,000,000,000}$. This number is a lot bigger than the 70-digit number we got for the car-parking problem. It's a lot bigger than L-space for normal-sized books, too. In fact, it has about 1,800,000,000 digits. If you wrote it out with 3,000 digits per page, you'd need a 600,000-page book to hold it.

The image of DNA-space is very useful for geneticists who are considering possible changes to DNA sequences, such as 'point mutations' where one code letter is changed, say as the result of a copying error. Or an incoming high-energy cosmic ray. Viruses, in particular, mutate so rapidly that it makes little sense to talk of a viral species as a fixed thing. Instead, biologists talk of quasi-species, and visualise these as clusters of related sequences in DNA-space. The clusters slosh around as time passes, but they stay together as one cluster, which allows the virus to retain its identity.

In the whole of human history, the total number of people has been no more than ten billion, a mere 11-digit number. This is an incredibly tiny fraction of all those possibilities. So actual human beings have explored the tiniest portion of DNA-space, just as actual books have explored the tiniest portion of L-space. Of course, the interesting questions are not as straightforward as that. Most sequences of letters do not make up a sensible book; most DNA sequences do not corre-spond to a viable organism, let alone a human being.

And now we come to the crunch for phase spaces. In physics, it is reasonable to assume that the sensible phase space can be 'pre-stated' before tackling questions about the corresponding system. We can imagine rearranging the bodies of the solar system into *any* config-uration in that imaginary phase space. We lack the engineering capacity to do that, but we have no difficulty imagining it done, and we see no physical reason to remove any particular configuration from consideration.

When it comes to DNA-space, however, the important questions are not about the whole of that vast space of all possible sequences. Nearly all of those sequences correspond to no organism whatsoever, not even a dead one. What we really need to consider is 'viable-DNA-space', the space of all DNA sequences that could be realised within some viable organism. This is some immensely complicated but very thin part of DNA-space, and we don't know what it is. We have no idea how to look at a hypothetical DNA sequence and decide whether it can occur in a viable organism.

The same problem arises in connection with L-space, but there's a twist. A literate human can look at a sequence of letters and spaces and decide whether it constitutes a story; they know how to 'read' the code and work out its meaning, if it's in a language they understand. They can even make a stab at deciding whether it's a good story or a bad one. However, we do not know how to transfer this ability to a computer. The rules that our minds use, to decide whether what we're reading is a story, are implicit in the networks of nerve cells in our brains. Nobody has yet been able to make those rules explicit. We don't know how to *characterise* the 'readable books' subset of L-space.

For DNA, the problem is compounded because there isn't some kind of fixed rule that 'translates' a DNA code into an organism. Biologists used to think there would be, and had high hopes of learning the 'lan-guage' involved. Then the DNA for a genuine (potential) organism would be a code sequence that told a coherent story of biological development, and all other DNA sequences would be gibberish. In effect, the biologists expected to be able to look at the DNA sequence of a tiger and *see* the bit that specified the stripes, the bit that specified the claws, and so on.

This was a bit optimistic. The current state of the art is that we can see the bit of DNA that specifies the protein from which claws are made, or the bits that make the orange, black and white pigments of the fur that show up as stripes, but that's about as far as our understanding of DNA narrative goes. It is now becoming clear that many non-genetic factors go into the growth of an organism, too, so even in principle there may not be a 'language' that translates DNA into living creatures. For example, tiger DNA turns into a baby tiger only in the presence of an egg, supplied by a mother tiger. The same DNA, in the presence of a mongoose egg, would not make a tiger at all.

Now, it could be that this is just a technical problem: that for each DNA code there is a unique kind of mother-organism that turns it into a living creature, so that the form of that creature is still *implicit* in the code. But theoretically, at least, the same DNA code could make two totally different organisms. We give an example in *The Collapse of Chaos*, where the developing organism first 'looks' to see what kind of mother it is in, and then develops in different ways depending on what it sees.

Complexity guru Stuart Kauffman has taken this difficulty a stage further. He points out that while in physics we can expect to pre-state the phase space of a system, the same is never true in biology. Biological systems are more creative than physical ones: the organisation of matter within living creatures is of a different qualitative nature from the organisation we find in inorganic matter. In particular organisms can evolve, and when they do that they often become more complicated. The fish-like ancestor of humans was less complicated than we are today, for example. (We've not specified a measure of complexity here, but that statement will be reasonable for most sensible measures of complexity, so let's not worry about definitions.) Evolution does not *necessarily* increase complexity, but it's at its most puzzling when it does.

Kauffman contrasts two systems. One is the traditional thermodynamic model in physics, of N gas molecules (modelled as hard spheres) bouncing around inside their $6N$-dimensional phase space. Here we know the phase space in advance, we can specify the dynamic precisely, and we can deduce general laws. Among them is the Second Law of Thermodynamics, which states that with over-whelming probability the system will become more disordered as time passes, and the molecules will distribute themselves uniformly through-out their container.

The second system is the 'biosphere', an evolving ecology. Here, it is not at all clear which phase space to use. Potential choices are either much too big, or much too limited. Suppose for a moment that the old biologists' dream of a DNA language for organisms was true. Then we might hope to employ DNA-space as our phase space.

However, as we've just seen, only a tiny, intricate subset of that space would really be of interest – but we can't work out which sub-set. When you add to that the probable non-existence of any such language, the whole approach falls apart. On the other hand, if the phase space is too small, entirely reasonable changes might take the organisms outside it altogether. For example, tiger-space might be defined in terms of the number of stripes on the big cat's body. But if one day a big cat evolves that has spots instead of stripes, there's no place for it in the tiger phase space. Sure, it's not a tiger ... but its mother was. We can't sensibly exclude this kind of innovation if we want to understand real biology.

As organisms evolve, they change. Sometimes evolution can be seen as the opening-up of a region of phase space that was sitting there waiting, but was not occupied by organisms. If the colours and pat-terns on an insect change a bit, all that we're seeing is the exploration of new regions of a fairly well-defined 'insect-space'. But when an entirely new trick, wings, appears, even the phase space seems to have changed.

It is very difficult to capture the phenomenon of innovation in a mathematical model. Mathematicians like to pre-state the space of pos-sibilities, but the whole point about innovation is that it opens up new possibilities that were previously not envisaged. So Kauffman suggests that a key feature of the biosphere is the inability to pre-state a phase space for it.

At risk of muddying the waters, it is worth observing that even in physics, pre-stating the phase space is not as straightforward as it might appear. What happens to the phase space of the solar system if we allow bodies to break up, or merge? Supposedly* the Moon was splashed off the Earth when it collided with a body about the size of Mars. Before that event, there was no Moon-coordinate in the phase space of the solar system; afterwards, there was. So the phase space expanded when the Moon came into being. The phase spaces of physics always assume a fixed context. In physics, you can usually get away with that assumption. In biology, you can't.

There's a second problem in physics, too. That $6N$-dimensional phase space of thermodynamics, for example, is too big. It includes non-physical states. By a quirk of mathematics, the laws of motion for elastic spheres do not prescribe what happens when three or more collide simultaneously. So we must excise from that nice, simple $6N$-dimensional space all configurations that experience a triple collision somewhere in their past or future. We know four things about these configurations. They are very rare. They can occur. They form an extremely complicated cloud of points in phase space. And it is impossible, in any practical sense, to determine whether a given configuration should or should not be excised. If these unphysical states were a bit more common, then the thermodynamic phase space would be just as hard to pre-state as that for the biosphere. However, they are a vanishingly small proportion of the whole, so we can just about get away with ignoring them.

Nonetheless, it *is* possible to go some way towards pre-stating a phase space for the biosphere. While we cannot pre-state a space of *all* possible organisms, we can look at any given organism and at least in principle say what the potential immediate changes are. That is, we can describe the space of the *adjacent* possible, the local phase space. Innovation then becomes the process of expanding into the adjacent possible. This is a reasonable and fairly conventional idea. But, more controversially, Kauffman suggests the exciting possibility that there may be general laws that govern this kind of expansion,

* See *The Science of Discworld*, 'A giant leap for moonkind'.

laws that have exactly the opposite effect to the famous Second Law of Thermodynamics. The Second Law in effect states that thermodynamic systems become simpler as time passes; all of the interesting structure gets 'smeared out' and disappears. In contrast, Kauffman's suggestion is that the biosphere expands into the space of the adjacent possible at the maximum rate that it can, subject to hanging together as a biological system. Innovation in biology happens *as rapidly as possible*.

More generally, Kauffman extends this idea to any system composed of 'autonomous agents'. An autonomous agent is a generalised lifeform, defined by two properties: it can reproduce, and it can carry out at least one thermodynamic work cycle. A work cycle occurs when a system does work and returns to its original state, ready to do the same again. That is, the system takes energy from its environment and transforms it into work, and does so in such a manner that at the end of the cycle it returns to its initial state.

A human being is an autonomous agent, and so is a tiger. A flame is not: flames reproduce by spreading to inflammable material nearby, but they do not carry out a work *cycle*. They turn chemical energy into fire, but once something has been burnt, it can't be burnt a second time.

This theory of autonomous agents is explicitly set in the context of phase spaces. Without such a concept, it cannot even be described. And in this theory we see the first possibility of obtaining a general understanding of the principles whereby, and wherefore, organisms complicate themselves. We are starting to pin down just what it is about lifeforms that makes them behave so differently from the boring prescription of the Second Law of Thermodynamics. We paint a picture of the universe as a source of ever-increasing complexity and organisation, instead of the exact opposite. We find out why we live in an interesting universe, instead of a dull one.

FIVE

REMARKABLY LIKE ANKH-MORPORK

'HOW CAN YOU COMMUNICATE LIKE THIS?' panted Ponder, as they jogged along beside a broad river.

'Since the physics of Roundworld are subordinate to the physics of the real world, I can use anything considered to be a communication device,' said the voice of Hex, slightly muffled in Rincewind's pocket. 'The owner of this device believes it to be one such. Also, I can deduce much information from this world's footprint in L-space. And the Archchancellor was right. There is much Elvish influence here.'

'You can extract information from Roundworld books?' said Ponder.

'Yes. The phase space of books that relate to this world contains ten to the power of 1,100 to the power of n volumes,' said Hex.

'That's enough books to fill the univ— hold on, what is n?'

'The number of all possible universes.'

'Then that's enough books to fill all possible universes! Well … as close as makes no difference, anyway.'

'Correct. That is why there is never enough bookshelf space. However, because of the subordinate temporal matrix of this world, I can use virtual computing,' said Hex. 'Once you know what the answer is, the process of calculation can be seriously reduced. Once the correct answer is found, the fruitless channels of inquiry cease to exist. Besides, if you deduct all the books that are about golf, cats, slood* and cookery the number is really quite manageable.'

* An extremely common and versatile substance, unfortunately not available in all universes.

'Oook,' said the Librarian.

'He says he's not going to have a shave,' said Rincewind.

'It is essential,' said Hex. 'We are getting strange glances from people in the fields. We do not wish to attract a mob. He must be shaved, and given a robe and hat.'

Rincewind was doubtful. 'I don't think that'll fool anyone,' he said.

'My readings tell me that it will if you say he's Spanish.'

'What's Spanish?'

'Spain is a country some five hundred miles from this one.'

'And people there look like him?'

'No. But people here would be quite prepared to believe so. This is a credulous age. The elves have done a lot of damage. The greatest minds spend half their time busying themselves with the study of magic, astrology, alchemy and communion with spirits.'

'Well? Sounds just like life at home,' said Rincewind.

'Yes,' said Hex. 'But there is no narrativium in this world. No magic. None of those things work.'

'Then why don't they just stop trying it?' said Ponder.

'My inference is that they believe it should work if only they get it right.'

'Poor devils,' said Rincewind.

'They believe in those, too.'

'There's more houses ahead,' said Ponder. 'We're coming to a city. Er ... and we've got the Luggage with us. Hex, we haven't just got an orangutan with us, we've got a box on legs!'

'Yes. We must leave it in some bushes while we find a voluminous dress and a wig,' said Hex calmly. 'Fortunately, this is the right period.'

'A dress won't work, believe me!'

'It will if the Librarian sits on the Luggage,' said Hex. 'That will bring him up to the right height and the dress will provide adequate cover for the Luggage.'

'Now hang on a moment,' said Rincewind. 'You saying that people here will believe an ape in a dress and a wig is a woman?'

'They will if you say she's Spanish.'

Rincewind took another look at the Librarian.

'Those elves really *must* have done a lot of damage,' he said.

The city was remarkably like Ankh-Morpork, although smaller and, unbelievably, smellier. One reason for that was the large number of animals in the streets. It was as if the place had been designed as a village and simply scaled up.

The wizards hadn't been hard to find. Hex located them easily, but in any case the noise could be heard in the next street. There was a tavern, with a courtyard, and in the courtyard a crowd of alcohol, which contained people, was watching a man trying to beat Archchancellor Ridcully with a very long and heavy staff.

He wasn't succeeding. Ridcully, who was stripped to the waist, was fighting back very effectively, putting his wizarding staff to the unusual task of hitting someone. He was a lot better at it than his opponent. Most wizards would die rather than take exercise, and did, but Ridcully had the rude health of a bear and only marginally better interpersonal skills. Despite his quite considerable if erratic erudition, at heart he was a man who'd rather smack someone around the ear than develop a complicated argument.

As the rescue party arrived, he hit the man across the head and then swept his feet from under him on the back-swing. A cheer went up as the man went down.

Ridcully helped his stunned adversary to his feet and propelled him to a bench, where the man's friends poured beer over him. Then he nodded to Rincewind and company.

'Got here, then,' he said. 'Bring the stuff, did you? Who's the Spanish lady?'

'That's the Librarian,' said Rincewind. There wasn't a great deal visible between the ruff and the red wig except an impression of extreme annoyance.

'Is it?' said Ridcully. 'Oh, yes. Sorry. Been here too long. This place gets to you. Good thinking, puttin' him in disguise. Hex suggested that, I expect.'

'We came as quick as we could, sir,' said Ponder. 'How long have you been here?'

'Couple of weeks,' said Ridcully. 'Not a bad place. Come and meet everyone.'

The rest of the wizards were sitting around a table. They were dressed in their normal wizarding outfits which, Rincewind had noticed, fitted in pretty well with the costumes in this town. But each man had equipped himself with a ruff, just to be on the safe side.

They nodded cheerfully at the newcomers. A forest of empty mugs in front of them went some way to explaining the cheer.

'You've detected elves?' said Ridcully, forcing enough wizards apart to give them seats.

'The place is lousy with glamour, sir,' said Ponder, sitting down.

'You're telling me,' said Ridcully. He glanced along the table. 'Oh, yes. We've found a friend. Dee, this is Mister Stibbons. Remember we told you about him?'

It was then that Ponder realised there were a couple of non-wizards in the party. It was quite hard to spot one, though, since for all practical purposes he fitted in well. He even had the right kind of beard.

'Er ... the noddlepate?' said Dee.

'No, that's Rincewind,' said Ridcully. 'Ponder is the *clever* one. And this ...' he turned to the Librarian, and words failed even him, 'is ... a ... friend of theirs.'

'From Spanish,' said Rincewind, who didn't know what noddlepate meant but had formed a pretty good idea.

'Dee here is a sort of local wizard,' said Ridcully, in the loud voice he thought was a confidential whisper. 'Sharp as a tack, mind like a razor, but spends all his spare time trying to do magic!'

'Which doesn't work here,' said Ponder.

'Right! But everyone believes it does, despite everything. Amazing! That's what elves can do to a place.' Ridcully leaned forward, conspiratorially. 'They came straight through our world and straight on into this one and we got caught up in the ... what's it you call it when it's all swirly and chilly as hell?'

'Trans-dimensional flux, sir,' said Ponder.

'Right. We'd have been totally lost if our friend Dee here hadn't been working a magic circle at the time.'

There was silence from Rincewind and Ponder. Then Rincewind said: 'You said magic doesn't work here.'

'As with this crystal sphere,' said a voice from Rincewind's pocket, 'this world is quite capable of maintaining a passive receptor.'

Rincewind removed the scrying stone from his pocket.

'But that is *mine*,' said Dee, staring at it.

'Sorry,' said Rincewind. 'We just sort of found it and sort of picked it sort of up.'

'But it *speaks*!' gasped Dee. 'An ethereal voice!'

'No, it's just from another world that is much bigger than this one and can't be seen,' said Ridcully. 'There's nothing mysterious about it at all.'

With trembling fingers, Dee took the sphere from Rincewind and held it in front of his eyes.

'Speak!' he commanded.

'Permission denied,' said the crystal. 'You do not have the rights to do this.'

'Where did you tell him you came from?' Rincewind whispered to Ridcully, as Dee tried to polish the ball with the sleeve of his robe.

'I just said we'd dropped in from another sphere,' said Ridcully. 'After all, this universe is full of spheres. He seemed to be quite happy about that. I didn't mention the Discworld at all, in case it confused him.'

Rincewind looked at Dee's shaking hands and the manic glint in his eye.

'I just want to be clear,' he said slowly. 'You appeared in a magic circle, you told him you're from another sphere, he'd just spoken to a crystal ball, you've explained to him that magic doesn't work and you don't want to confuse him?'

'Make him any more confused than he is already, you mean,' said the Dean. 'Confusion is the natural state of mind here, believe us. Do you know they think numbers are magical? Doing sums can get a man into real trouble in these parts.'

'Well, some numbers *are* magi—' Ponder began.

'Not here they're not,' said the Archchancellor. 'Here I am, out in the open air, no magical protection and I'm going to say the number that comes after seven. Here it comes: *eight*. There. Nothing happened. Eight! Eighteen! Two fat ladies in very tight corsetry, eighty-eight! Oh, someone pull Rincewind out from under the table, will you?'

While the Professor of Cruel and Unusual Geography was having some of it brushed off his robes, Ridcully continued: 'It's a mad world. No narrativium. People makin' up history as they go along. Brilliant men spendin' their time wondering how many angels can dance on the head of a pin—'

'Sixteen,' said Ponder.

'Yes, *we* know that because *we* can go and look, but here it's just another silly question,' said Ridcully. 'It'd make you cry. The history of this place goes backwards half the time. It's a mess. A parody of a world.'

'We made it,' said the Lecturer in Recent Runes.

'We didn't make it *this* badly,' said the Dean. 'We've seen the history books here. There were some great civilisations thousands of years ago. There was a place like Ephebe that was really beginning to find things out. The wrong things, mostly, but at least they were making an effort. Even had a decent pantheon of gods. All gone now. Our chum here and his friends think everything worth knowing has been discovered and forgotten and, frankly, they're not totally wrong.'

'What can we do about it?' said Ponder.

'You can talk to Hex on that thing?'

'Yes, sir.'

'Then Hex can do the magic back at UU and we'll find out what the elves did,' said Ridcully.

'Er,' Rincewind began, 'do we have the *right* to interfere?' They all stared at him.

'I mean, we never did it before,' he went on. 'Remember all those other creatures that evolved here? The intelligent lizards? The intelligent crabs? Those dog things? They all got completely wiped out by ice ages and falling rocks and we never did anything to stop it.'*

They went on staring.

'I mean, elves are just another problem, aren't they?' said Rincewind. 'Maybe … maybe they're just another form of big rock? Maybe … maybe they always turn up when intelligence gets going? And the species is either clever enough to survive them or it ends up buried

* The sad histories of these hitherto unknown civilisations, along with the tale of the two-mile limpet, can be found in *The Science of Discworld*.

in the bedrock like all the others? I mean, perhaps it's a kind of, a kind of test? I mean …'

It dawned on Rincewind that he was not carrying the meeting. The wizards were glaring at him.

'Are you suggesting that someone somewhere is awarding *marks*, Rincewind?' said Ponder.

'Well, obviously there is no—'

'Good. Shut up,' said Ridcully. 'Now, lads, let's get back to Mortlake and get started.'

'Mort Lake?' said Rincewind. 'But that's in Ankh-Morpork!'

'There's one here, too,' said the Lecturer in Recent Runes, beaming. 'Amazing, isn't it? We never guessed. This world is a cheap parody of our own. As Above, So Below and all that.'

'But without magic,' said Ridcully. 'And with no narrativium. It doesn't know where it's going.'

'But *we* do, sir,' said Ponder, who had been scribbling in his notebook.

'Do we?'

'Yes, sir. Remember? In about a thousand years' time it's going to be hit by a really big rock. I keep looking at the numbers, sir, and that's what it means.'

'But I thought we found there'd been a race that built huge structures to get off the place?'

'That's right, sir.'

'Can a new species turn up in a thousand years?'

'I don't think so, sir.'

'You mean *these* are the ones that leave?'

'It seems like it, sir,' said Ponder.

The wizards looked at the people in the courtyard. Of course, the presence of beer always greases the rungs of the evolutionary ladder, but even so …

At a nearby table, one man threw up on another one. There was general applause.

'I think,' said Ridcully, summing up the general mood, 'that we are going to be here for some time.'

SIX

THE LENS-GRINDER'S PHILOSOPHY

 JOHN DEE, WHO LIVED from 1527 to 1608, was court astrologer to Mary Tudor. At one point he was imprisoned for being a magician, but in 1555 they let him out again, presumably for not being one. Then he became astrologer for Queen Elizabeth I. He devoted much of his life to the occult, both alchemy and astrology. On the other hand, he was also the author of the first English translation of Euclid's *Elements*, the renowned treatise on geometry. Actually, if you believe the printed word, the book is attributed to Sir Henry Billingsley, but it was common knowledge that Dee did all the work, and he even wrote a long and erudite preface. Which may be *why* it was common knowledge that Dee did all the work.

To the modern mind, Dee's interests seem contradictory: a mass of superstitious pseudoscience mixed up with some good, solid science and mathematics. But Dee didn't have a modern mind, and he saw no particular contradiction in the combination. In his day, many mathematicians made their living by casting horoscopes. They could do the sums that foretold in which of the twelve 'houses' – the regions of the sky determined by the constellations corresponding to the signs of the zodiac – a planet would be.

Dee stands at the threshold of modern ways of thinking about causality in the world. We call his time The Renaissance, and the reference is to the rebirth of the philosophy and politics of ancient Athens. But perhaps this view of his times is mistaken, both because Greek society was not then as 'scientific' or 'intellectual' as we've been led to believe, and because there were other cultural currents that contributed to the culture of his times. Our ideas of narrativium may derive from the melding of these ideas into later philosophies, such as that of Baruch Spinoza.

Stories encouraged the growth of occultism and mysticism. But they also helped to ease the European world out of medieval superstition into a more rational view of the universe.

Belief in the occult – magic, astrology, divination, witchcraft, alchemy – is common to most human societies. The European tradition of occultism, to which Dee belonged, is based on an ancient, secret philosophy; it derives from two main sources, ancient Greek alchemy and magic, and Jewish mysticism. Among the Greek sources is the *Emerald Tablet*, a collection of writings associated with Hermes Trismegistos ('thrice master'), which was particularly revered by later Arab alchemists; the Jewish source is the *Kabbala*, a secret, mystical interpretation of a sacred book, the *Torah*.

Astrology, of course, is a form of divination based on the stars and the visible planets. It may, perhaps, have contributed to the development of science by supporting people who wanted to observe and understand the heavens. Johannes Kepler, who discovered that planetary orbits are ellipses, made his living as an astrologer. Astrology still survives in watered-down form in the horoscope columns of tabloid newspapers. Ronald Reagan consulted an astrologer during his time as American President. That stuff certainly hangs around.

Alchemy is more interesting. It is often said to be an early forerunner of chemistry, although the principles underlying chemistry largely derive from other sources. The alchemists played around with apparatus that led to useful chemists' gadgets like retorts and flasks, and they discovered that interesting things happen when you heat certain substances or combine them together. The alchemists' big discoveries were sal ammoniac (ammonium chloride), which can be made to react with metals, and the mineral acids – nitric, sulphuric and hydrochloric.

The big goal of alchemy would have been much bigger if they'd ever achieved it: the Elixir of Life, the source of immortality. The

Chinese alchemists described this long-sought substance as 'liquid gold'. The narrative thread here is clear: gold is the noble metal, incorruptible, ageless. So anyone who could somehow incorporate gold into his body would also become incorruptible and ageless. The nobility shows up differently: the noble metal is reserved for the 'noble' humans: emperors, royalty, the people on top of the heap. Much good did this do them. According to the Chinese scholar Joseph Needham, several Chinese emperors probably died of elixir poisoning. Since arsenic and mercury were common constituents of supposed elixirs, this is hardly a surprise. And it is all too plausible that a mystic quest for immortality would shorten life, not prolong it.

In Europe, from about 1300 onwards, alchemy had three main objectives. The Elixir of Life was still one, and a second was finding cures for various diseases. The alchemical search for medicines eventually led somewhere useful. The key figure here is Phillipus Aureolus Theophrastus Bombastus* von Hohenheim, mercifully known as 'Paracelsus', who lived from 1493 to 1541.

Paracelsus was a Swiss physician whose interest in alchemy led him to invent chemotherapy. He placed great store in the occult. As a student aged 14 he wandered from one European university to another, in search of great teachers, but we can deduce from what he wrote about the experience, somewhat later, that he was disappointed. He wondered why 'The high colleges managed to produce so many high asses', and clearly wasn't the kind of student to endear himself to his teachers. 'The universities,' he wrote, 'do not teach all things. So a doctor must seek out old wives, gypsies, sorcerers, wandering tribes, old robbers, and such outlaws and take lessons from them.' He would have had a high old time on Discworld, but would have learned a lot.

After ten years' wandering, he returned home in 1524 and became lecturer in medicine at the university of Basel. In 1527 he publicly burned the classic books of earlier physicians, the Arab Avicenna and the Greek Galen. Paracelsus cared not a whit for authority. Indeed his assumed name, 'para-Celsus', means 'above Celsus', and Celsus was a leading Roman doctor of the first century.

---

* Isn't 'Bombastus' a lovely name? Well-chosen, too.

He was arrogant and mystical. His saving grace was that he was also very bright. He placed great importance on using nature's own powers of healing. For example, letting wounds drain instead of padding them with moss or dried dung. He discovered that mercury was an effective treatment for syphilis, and his clinical description of that sexually transmitted disease was the best available.

The main objective for most alchemists was far more selfish. Their sights were set on just one thing: transmuting base metals like lead into gold. Again, their belief that this was possible rested on a story. They knew from their experiments that sal ammoniac and other substances could change the colour of metals, so the story 'Metals can be transmuted' gained ground. Why, then, should it not be possible to start with lead, add the right substance, and end up with gold? The story seemed compelling; all that they lacked was the right substance. They called it the Philosopher's Stone.

The search for the Philosopher's Stone, or rumours that it had been found, got several alchemists into trouble. Noble gold was the prerogative of the nobility. While the various kings and princes wouldn't have minded getting their hands on an inexhaustible supply of gold, they didn't want their rivals to beat them to it. Even *searching* for the Philosopher's Stone could be considered subversive, just as searching for a cheap source of renewable energy now is apparently considered subversive by oil corporations and nuclear energy companies. In 1595 Dee's companion Edward Kelley was imprisoned by Rudolf II and died trying to escape, and in 1603 Christian II of Saxony imprisoned and tortured the Scottish alchemist Alexander Seton. A dangerous thing, a clever man.

The story of the Philosopher's Stone never reached its climax. The alchemists never did turn lead into gold. But the story took a long time to die. Even around 1700, Isaac Newton still thought it was worth having a go, and the idea of turning lead into gold by *chemical* means was finally killed off only in the nineteenth century. Nuclear reactions, mind you, are another matter: the transmutation can be done, but it is wildly uneconomic. And unless you're very careful, the gold is radioactive (although, of course, this will keep the money circulating quickly, and we might see a sudden upsurge of philanthropy).

How did we get from alchemy to radioactivity? The pivotal period of Western history was the Renaissance, roughly spanning the fifteenth and sixteenth centuries, when ideas imported from the Arab world collided with Greek philosophy and mathematics, and Roman artisanship and engineering, leading to a sudden flowering of the arts and the birth of what we now call science. During the Renaissance, we learned to tell new stories about ourselves and the world. And those stories changed both.

In order to understand how this happened, we must come to grips with the real Renaissance mentality, not the popular view of a 'Renaissance man'. By that phrase, we mean a person with expertise in many areas – like Roundworld's Leonardo da Vinci, who bears a suspicious resemblance to the Disc's Leonard of Quirm. We use this phrase because we contrast such people with what we call a 'well-educated' person today.

In medieval Europe, and indeed long after that, the aristocracy considered 'education' to mean classical knowledge – the culture of the Greeks – plus a lot of religion, and not much else. The king was expected to be well informed about poetry, drama and philosophy, but he wasn't expected to know about plumbing or brickwork. Some kings did in fact get rather interested in astronomy and science, either out of intellectual interest or the realisation that technology is power, but that wasn't part of the normal royal curriculum.

This view of education implied that the classics were all the validated knowledge that an 'educated' person needed, a view not far from that of many English 'public' schools until quite recently, and of the politicians they have produced. This view of what was needed by the rulers contrasted with what was needed by the children of the peasantry (artisan skills and, lately, the 'three Rs'*).

Neither the classics nor the three Rs formed the basis for the genuine Renaissance man, who sought a fusion of those two worlds. Pointing to the artisan as a source of worldly experience, of knowledge

---

* Readers who have not met this felicitous phrase, for reasons of youth or geography, should be told that the three Rs are Reading, Riting and Rithmetic. What this tells us about the educational establishment is unclear, but it *could* be a joke. The three Rs, not the educational establishment, that is. Though, come to think of it …

of the material world and its tools such as an alchemist might use, led to a new rapprochement between the classical and the empirical, between intellect and experience. The actions of such men as Dee – even those of the occultist Paracelsus in his medical prescriptions – emphasised this distinction, and started the fusion of reason and empiricism that so impresses us today.

As we've said, the word 'Renaissance' refers not just to rebirth, but to a specific rebirth, that of ancient Greek culture. This, however, is a modern view, based on a mistaken view of the Greeks, and of the Renaissance itself. In 'classical' education, no attention is paid to engineering. Of course not. Greek culture ran on pure intellect, poetry and philosophy. They didn't have engineers.

Oh, but they did. Archimedes constructed great cranes that could lift enemy ships out of the water, and we still don't know exactly how he did it. Hero of Alexandria (roughly contemporary with Jesus) wrote many texts about engines and machines of various kinds of the previous three hundred years, many of which show that prototypes must have been made. His coin-operated machines were not too different from those that could be found on any city street in 1930s London or New York, and would probably have been more reliable when it came to disgorging the chocolate, if the Greeks had known about chocolate. The Greeks had elevators, too.

The problem here is that information about the technical aspects of Greek society has been transmitted to us through a bunch of theologians. They liked Hero's steam engine, and indeed many of them had a little glass one on their desk, a sort of Theologians' Toy that they could spin with a candle flame. But the mechanical ideas behind such toys just passed them by. And, just as Greek engineering has not been transmitted to us by theologians, the spiritual attitude of the Renaissance has not come down to us through our 'rational' schoolteachers. Much of the attempted spirituality within the alchemical position was basically a religious stance, marvelling at the Works of the Lord as they were exposed by the marvels of changes of state and form, when materials were subjected to heat, to 'percussion', and to solution and crystallisation.

This stance has been taken over by today's innocents of rigorous thinking, the New Agers, who find spiritual inspiration in crystals and anodised metals, spherical spark-machines and Newton's pendulums, but do not ask the deeper questions that lie behind these toys. We find the very real awe inspired by science's quest for understanding to be considerably more spiritual than New Age attitudes.

Today there are mystic massage-therapists, aromatherapists, iridologists, people who believe that you can 'holistically' tell what's wrong with someone by examining their irises or the balls of their feet – only – and who root their beliefs in the writings of Renaissance eccentrics like Paracelsus and Dee. But those men would have been horrified to be cited as authorities, especially by such closed-minded descendants.

Prominent among those who refer back to Paracelsus for authority are homeopathists. A basic belief of homeopathy is that medicines become more powerful the more they are diluted. This stance lets them promote their medicine as being totally harmless (it's just water) but also extraordinarily effective (as water isn't). They notice no contradiction here. And homeopathic headache tablets say 'Take one if mild, three if painful'. Shouldn't it be the other way round?

Such people see no need to think about what they are doing, because they base their beliefs on authority. If a question is not raised by that authority, then it's not a question they want to ask. So, in support of their theories, homeopaths quote Paracelsus: 'That which makes disease is also the cure.' But Paracelsus built his entire career on *not* respecting authority. Moreover, he never said that a disease is *always* its own cure.

Contrast this modern spectrum of silliness with the robust, critical attitude of most Renaissance scholars to the idea that arcane practices can lay bare the bones of the world. People such as Dee, indeed Isaac Newton, took that critical position very seriously. To a great extent, so did Paracelsus: for example he repudiated the idea that the stars and planets control various parts of the human body. The Renaissance view was that God's creation has mysterious elements, but those elements are hidden,* implicit in the nature of the universe, rather than arcane.

---

* Hidden knowledge at that time was spectacularly practical knowledge, exemplified by the Guild secrets and especially by the Freemasons. It was dressed up in ritual, because it was mostly passed on verbally and not written down.

This view is very close to Antonie van Leeuwenhoek's marvelling at the animalcules in dirty water, or semen: the astonishing discovery that the Wonders of Creation extended down into the microscopic realm. Nature, God's Creation, was much more subtle. It provided hidden wonders to marvel at as well as the overt artistic vision. Newton was taken with the implicit mathematics of the planets in just this way: there was more to God's invention than was apparent to the unaided eye, and that resonated with his Hermetic beliefs (a philosophy derived from the ideas of Hermes Trismegistos). The crisis of atomism at the time was the crisis of pre-formation: if Eve had within her all her daughters, each having within her *her* daughters like a set of Russian dolls, then matter must be infinitely divisible. Or, if not, we could work out the future date of Judgement Day by discovering how many generations there were until we got to the last, empty daughter.

A characteristic of Renaissance thinking, then, was a degree of humility. It was critical about its own explanations. This attitude contrasts favourably with such modern religions as homeopathy or scientology, creeds that arrogantly claim to offer a 'complete' explanation of the Universe in human terms.

Some scientists are equally arrogant, but good scientists are always aware that science has limitations, and are willing to explain what they are. 'I don't know' is one of the great, though admittedly under-utilised, scientific principles. Admitting ignorance clears away so much pointless nonsense. It lets us cope with stage magicians performing their beautiful, and very convincing, illusions – convincing, that is, while we keep our brains out of gear. We know they have to be tricks, and admitting ignorance lets us avoid the trap of believing the illusion to be real merely because we don't know how the trick works. Why should we? We're not members of the Magic Circle. Admitting ignorance similarly protects us against mystic credulity when we encounter natural events that have not yet caught the eye of a competent scientist (and his grant-awarding body), and that still seem to be … magic. We say 'The magic of nature' … more the Wonder of Nature, or the Miracle of Life.

This is a stance that nearly all of us share, but it's important to understand the historical tradition it is grounded in. It isn't simply a case of admiring the complexity of God's works. It implies the attitudes of Newton, van Leeuwenhoek and earlier; indeed, right back to Dee. And, doubtless, to some Greek, or several. It involves the Renaissance belief that if we investigate the wonder, the marvel, the miracle, then we'll find even more wonders, marvels and miracles: gravity, say, or spermatozoa.

So what do we, and what did they, mean by 'magic'? Dee spoke of the arcane arts, and Newton was committed to many explanations that were 'magickal', especially his commitment to action at a distance, 'gravity', which derived from the mystical attraction/repulsion basics of his Hermetic philosophy.

So 'magic' means three things, all apparently quite different. Meaning one is: 'something to be wondered at', and this ranges from card tricks to amoebas to the rings of Saturn. Meaning two is turning a verbal instruction, a *spell*, into material action, by occult or arcane means ... turning a person into a frog, or *vice versa*, or a *djinn* building a castle for his master. The third meaning is the one we use: the technical magic of turning a light switch on, and getting light, without even having to say 'fiat lux'.

Granny Weatherwax's recalcitrant broomstick is type two magic, but her 'headology' is largely a very, very good grasp of psychology (type three magic carefully disguised as type two). It brings to mind Arthur C. Clarke's phrase 'Any sufficiently advanced technology is indistinguishable from magic', which we quoted and discussed in *The Science of Discworld*. Discworld exemplifies magic by spells, and indeed is maintained as an unlikely creation by being immersed in a strong magical field (type two). Adults of Earthly cultures, like Roundworld, pretend to have lost intellectual belief in magic of the Discworld kind, while their culture is turning more and more of their technology into magic (third kind). And the development of Hex throughout the books is turning Sir Arthur on his head: Discworld's sufficiently advanced magic is now practically indistinguishable from technology.

We can see, as (fairly) rational adults, where the first kind of magic comes from. We see something wonderful and feel tremendously happy that the universe is a place that can include ammonites, say, or kingfishers. But where did we get our belief in the second, irrational kind of magic? How does it come about that all cultures have children that begin their intellectual lives by believing in magic, instead of the real causality that surrounds them?

A plausible explanation is that human beings are initially programmed through fairy stories and nursery tales. All human cultures tell stories to their children; part of the development of our specific humanity is the interaction that we get with early language.

All cultures use animal icons for this nursery tuition, so we in the West have sly foxes, wise owls and frightened chickens. They seem to come out of a human dreamtime, where all animals seem to be types of human being in a different skin, and talk as a matter of course. We learn what the subtle adjectives mean from the actions – and words – of the creatures in the stories. Inuit children don't have a 'sly' fox icon; their fox is 'brave' and 'fast', while the Norwegian iconic fox is secretive and wise, full of good advice for respectful children. The causality in these stories is always verbal: 'So the fox said ... and they did it!' or 'I'll huff and I'll puff and I'll blow your house down.' The earliest communicated causality that the child meets is verbal instructions that cause material events. That is, spells.

Similarly, parents and carers are always transmuting the child's expressed desires into actions and objects, from food appearing on the table when the child is hungry to toys and other birthday and Christmas gifts. We surround these simple verbal requests with 'magical' ritual. We require the spell to begin with 'please', and its execution to be recognised by 'thank you'.* It is indeed not surprising that our children come to believe that the way to acquire or access bits of the real world is simply to ask – indeed, simply asking or commanding is *the* classic spell. Remember 'open, sesame'?

To a child, the world does work like magic. Later in life, we wish that we could go on like that, with our 'wishes coming true'.† So we

---

* Carers even encourage or berate the child: 'What's the magic word? You forgot the magic word!'

† Years ago, Jack wrote a book called *The Privileged Ape* about just this tendency. What he wanted to call it – and should have, but the publisher got cold feet – was *The Ape That Got What It Wanted*. (When it gets it, of course, it no longer wants it.)

design our shops, our webpages, our cars to fit this truly 'childish' view of the world.

Coming home in the car and clicking the garage open, clicking the infrared remote to open or lock the car, changing TV channels – even switching on the light by the wall switch – are just that kind of magic. Unlike our Victorian forebears, we like to hide the machinery and pretend it's not there. So Clarke's dictum is not at all surprising. What it means is that this ape keeps trying, with incredible ingenuity, to get back into the nursery, when everything was done for it. Maybe other intelligent/extelligent species will have a similar helpless early life, which they will attempt to compensate for or relive through their technology? If so, they will 'believe in magic', too, and we will be able to diagnose this by their possession of 'please' and 'thank you' rituals.

We can see this philosophy surviving into adulthood in different human cultures. In 'adult' stories like the Arabian Nights, an assortment of *djinni* and other marvels grant the heroes' wishes by magical means, just like those child-wishes coming true. Many 'romantic' adult stories have the same kind of setting, as do many fantasy tales. Fairness demands we add that, contrary to popular opinion, modern fantasy stories don't; it's hard to get much tension in a plot when anything is possible at the snap of a wand and so the practice of 'magic' therein tends to be difficult, dangerous and to be avoided wherever possible. Discworld is a magical world – we can hear the thoughts of a thunderstorm, for example, or the conversation of dogs – but magic in the pointy hat sense is very seldom used. The wizards and witches treat it rather like nuclear weaponry: it does no harm for people to know you've got it, but *everyone* will be in trouble if it gets used. This is magic for grown-ups; it has to be hard, because we know there's no such thing as a free goblin.

Unfortunately, adult beliefs about causality are usually contaminated by the less sophisticated wish-fulfilment philosophy that we carry with us from the tinkly magic of our infancies. For example, scientists will object to alternative theories on the grounds that 'if that was true, we wouldn't be able to do the sums'. Why do they think that nature cares whether humans can do the sums? Because their own desire to do the sums, which lets them write papers for learned journals, contaminates their otherwise rational view. There's a feeling of feet being stamped; the Almighty should change Her laws so that we *can* do the sums.

There are other ways to set up beliefs about causality, but they are difficult for creatures immersed in their own cultural assumptions: nearly everything that an adult human being is required to do is either made magical by technology, or it is to do with another human being, serving or being served.

These management, leadership and aristocracy issues have been handled very differently in different societies. Feudal societies have a baronial class, who are in many respects allowed to remain in their nursery personas by being surrounded by servants and slaves and other parent-surrogates. Rich people in more complex societies, and high-status people in general (knights, kings, queens, princesses, Mafia bosses, operatic divas, pop idols, sports stars) seem to have set up societies around them that pander to their needs in a very child-pampering way. As our society has become more technical, more and more of us, right down to the lowest status levels of society, have come to benefit from the accumulating magic of technology. Supermarkets have democratised and validated the provision of all we could want to each of our child-natures. The child-magic has been appropriated by more and more adults, through technology, and the legitimate kind, the 'wonder of nature' magic, has lost out.

In the mid-seventeenth century there was a philosopher, Baruch Spinoza, who derived from the synthetic Renaissance position, and from his criticism of Descartes' publications, a wholly new view of causality. He was one of several figures who bridged the Renaissance and helped engender the Enlightenment. He developed his critical view of his own Jewish cultural authorities into a new rational view of universal causality. He rejected Moses' hearing God's voice, and angels, and lots more 'occult' thinking, particularly early cabbalism;* he took the naïve magic out of his own religion. He was a lens-grinder, an occupation that requires the persistent checking of performance against reality. So he put in the artisan's view of causality, and he took

* A system of mystic beliefs based on the Jewish *Kabbala*.

out the magic of God's word. The Jewish community in Amsterdam excommunicated him. They'd learned about that from the Catholics, but it didn't translate very well into Jewish practice, even of those times.

Spinoza was a pantheist. That is, he believed there is a little bit of God in everything. His main reason for believing this was that if God were separate from the material universe, then there would be an entity greater than God, namely, the entire universe plus God. It follows that Spinoza's God is not a *being*, not a person in whose image humanity can be made. For this reason, Spinoza was often considered to be an atheist, and many orthodox Jews still view him that way. Despite this, his *Ethics* makes a beautiful, logically argued case for a particular type of pantheism. In fact, Spinoza's viewpoint is almost indistinguishable from that of most philosophically inclined scientists, from Newton to Kauffman.

Before Spinoza, even his supposed predecessors like Descartes and Leibniz had God moving things in the World by the power of his Voice: magic, child-thinking. Spinoza introduced the idea that an overarching God could run the universe without being anthropomorphic. Many modern Spinozans see the set of rules, devised, described or attributed by science to the physical world, as the embodiment of that kind of God. That is to say, what happens in the material world happens that way because God, or the Nature of the Physical World, constrains it to do so. And out of that come ideas resembling narrativium instead of magic and wish-fulfilment.

A Spinozan view of child development sees the opposite of wish-fulfilment. There are rules, constraints, that *limit* what we can do. The child learns, as she grows, to modify her plans as she perceives more of the rules. Initially, she might attempt to cross the room assuming that the chair is not an obstacle; when it doesn't move out of her way, she will feel frustration, a 'passion'. And throws a paddy. Later, as she constructs her path to avoid the chair, more of her plans will peaceably, and successfully, come to fruition. As she grows and learns more of the rules – God's Will or the warp and woof of universal causation – this progressive success will produce a calm acceptance of constraints: peace rather than passion.

Kauffman's *At Home in the Universe* is a very Spinozan book, because Spinoza saw that we do indeed make our home, with the reward of peace and the discipline of passion and its control, each of us in their own universe. We fit the universe as a whole, we evolved in it and of it, and a successful life is based on appreciating how it constrains our plans and rewards our understanding. 'Please' and 'thank you' have no place in Spinozan prayer. That view melds the artisan with the philosopher, the tribal respect for tradition with the barbarian virtues of love and honour.

And it gives us a wholly new kind of story with a civilising message. Instead of the barbarian 'And then he rubbed the lamp again ... and again the genie appeared', we have the first king's son taking on a task, to win the hand of the fair princess ... and he fails. Amazing! No barbarian protagonist ever fails. Indeed, nobody *ever* ultimately fails except evil giants, sorcerers and Grand Viziers, in tribal or barbarian magical tales. However, the new story tells of the second king's son learning from this failure, and shows the listener – the learner – how difficult the task is. Nevertheless, *again he fails*, because learning is not easy. But the third son – or the third billygoat Gruff or the third pig, with his house of brick – shows how to succeed in a Spinozan, enlightened world of observation and experience. Stories in which people learn from the failures of others are a hallmark of a civilised society.

Narrativium has entered our Make-a-Human kit. It makes a different kind of mind from the tribal one, which is all 'do this because we've always done it that way and it works' and 'don't do that because it's taboo, evil and we'll kill you if you do'. And it also differs from the barbarian mind: 'That way lies honour, booty, much wealth and many children (if I can only get a *djinn*, or a dgun); I would not demean myself, dishonour these hands, with menial work.' In contrast, the civilised child learns to repeat the task, to work with the grain of the universe.

The reader of tales that have been moulded and informed by narrativium is prepared to do whatever an understanding of the task requires. Perhaps, in the universe of the story, qualifying for princesses' hands in marriage isn't the preoccupation of the average middle-class, but the attitude of the third prince will serve him well down the mine, in the Stock Exchange, in the Wild West (according to Hollywood, a great purveyor of narrativium), or as father and baron. We say 'he' because 'she' has a more difficult time: narrativium has not been mined and modelled for girls, and the way the feminist myths are shaping it does not seem to address the same questions as the old boy-oriented models. But we can put that right if we realise that narrativium trains by constraint.

Discworld, although technically a world run on fairy tale rules, derives much of its power and success from the fact that they are consistently challenged and subverted, most directly by the witch Granny Weatherwax, who cynically uses them or defies them as she sees fit. She roundly objects to girls being forced by the all-devouring 'story' to marry a handsome prince solely on the basis of their shoe size; she believes that stories are there to be challenged. But she herself is part of a larger story, and they follow rules, too. In a sense, she's always trying to saw off the branch she's sitting on. And her stories derive their power from the fact that we have been programmed from an early age to believe in the monsters that she is battling.

SEVEN

CARGO CULT
MAGIC

 THE PHRASE THAT KEPT OCCURRING to Rincewind was *cargo cult*.

He'd run across it – he encountered most things by running across them – on isolated islands out on the big oceans.

Say that, once, a lost ship arrived, and while taking on food and water it handed out a few goodies to the helpful locals, like steel knives, arrowheads and fish-hooks.* And then it sailed away, and after a while the steel wore out and the arrowheads got lost.

What was needed was another ship. But not many ships came to these lonely islands. What was needed was a *ship attractor*. Some sort of *decoy*. And it didn't much matter if it was made out of bamboo and palm leaves, so long as it looked like a ship. Ships would be bound to be attracted to another ship, or else how did you get small boats?

As with many human activities, it made perfect sense, for certain values of 'sense'.

Discworld magic was all about controlling the vast oceans of magic that poured though the world. All the Roundworld magicians could do was to build something like bamboo decoys on the shores of the big, cold, spinning universe, which pleaded: please let the magic come.

'It's terrible,' he said to Ponder, who was drawing a big circle on

---

* And new diseases, although it was quite hard to make bamboo models of these.

the floor, to Dee's fascination. 'They believe they live in *our* world. With the turtle and everything!'

'Yes, and that's strange because the rules here are quite easy to spot,' said Ponder. 'Things tend to become balls, and balls tend to move in circles. Once you work that out, everything else falls into place. In a curved movement, of course.'

He went back to chalking the circle.

The wizards had been staying in Dee's house. He seemed quite happy about this, in a mildly bemused way, like a peasant who had suddenly been visited by a family of unexpected relatives from the big city who were doing incomprehensible things but were rich and interesting.

The trouble was, Rincewind thought, that the wizards were explaining to Dee that magic didn't work while, at the same time, doing magic. A crystal ball was giving instructions. An ape was knuckling in and out of, for want of a better word, fresh air, and wandering around Dee's library making excited 'ook' noises and assembling the books to make a proper entrance in L-space. And the wizards themselves, as was their wont, prodded at things and argued at cross-purposes.

And Hex had tracked down the elves. It made no sense, but their descent on Roundworld had plunged through time and come to rest millions of years in the past.

Now the wizards had to get there. As Ponder explained, sometimes resorting to hand gestures for the hard of comprehension, this wasn't difficult. Time and space in the round universe were entirely subordinate. The wizards, being made of higher-order stuff, could quite easily be moved around within it by magic from the *real* world. There were additional, complex reasons, mostly quite hard to spell.

The wizards didn't understand almost all this, but they did like the idea of being high-order stuff.

'But there was *nothing* back there,' said the Dean, watching Ponder work on the circle. 'There wasn't even anyone you could call people, Hex says.'

'There were monkeys,' said Rincewind. 'Things like monkeys, any-way.' He had his own thoughts on this score, although the accepted wisdom on Discworld was that monkeys were the descendants of people who had given up trying.*

'Oh, the *monkeys*,' snapped Ridcully. 'I remember them. Completely useless. If you couldn't eat it or have sex with it, they just didn't want to know. They just mucked about.'

'I think this was even before that,' said Ponder. He stood up and brushed chalk dust off his robe. 'Hex thinks that the elves did something to … something. Something that became humans.'

'Interfered with them?' said the Dean.

'Yes, sir. We know they can affect people's minds when they sing—'

'You said *became humans*?' said Ridcully.

'Yes, sir. Sorry, sir. I really don't want to have that argument all over again, sir. On Roundworld, things become other things. At least, *some* of some things become other things. I'm not saying that happens on Discworld, sir, but Hex is quite certain that it happens here. Can we just pretend for a moment, sir, that this is true?'

'For the sake of argument?'

'Well, for the sake of not having an argument, sir, really,' said Ponder. Mustrum Ridcully on the subject of evolution could go on for far too long.

'All right, then,' said the Archchancellor with some reluctance.

'And we know, sir, that elves can *really* affect the minds of lesser creatures …'

Rincewind let the words go over his head. He didn't need to be told this. He'd spent far more time in the field – and the ditch, the forest, hiding in the reeds, staggering across deserts – and had run into and away from elves a couple of times. They didn't like all the things that Rincewind thought made life worthwhile, like cities and cookery and not being hit over the head with rocks on a regular basis. He'd never been certain if they actually *ate* anything, other than for amusement; they acted as if what they really consumed was other creatures' fear.

They must have loved humanity when they found it. Humanity was

* The Librarian, on the other knuckly hand, held the view that humans were *apes* who had given up trying. They were the ones who simply couldn't cut the mustard when it came to living in harmony with their environment, maintaining a workable social structure and, above all, sleeping while holding on.

very creative, when it came to being frightened. It was *good* at filling the future full of dread.

And then it had gone and spoiled everything by using that wonderful, fear-generating mind for thinking up things to take the fear away – like calendars, locks, candles and stories. Stories in particular. Stories were where the monsters died.

While the wizards argued, Rincewind went to see what the Librarian was doing. The ape, shorn of his dress but still wearing his ruff to conform to local clothing standards, was as happy as, well, as happy as a librarian among books. Dee was quite a collector. Most of the books were about magic or numbers or magic *and* numbers. They weren't very magical, though. The pages didn't even turn by themselves.

The crystal sphere had been placed on a shelf, so that Hex could watch.

'The Archchancellor wants us all to go back and stop the elves,' said Rincewind, sitting down on a stack of titles. 'He thinks we can ambush them before they do anything. Me, I don't think it's going to work.'

'Ook?' said the Librarian, sniffing a bestiary and laying it aside.

'Because things generally don't, that's why. Best laid plans, and all that. And these aren't best laid plans, anyway. "Let's get back there and beat the devils to death with big iron bars" is not, in my opinion, a best laid plan. What's funny?'

The Librarian's shoulders were shaking. He passed a book across to Rincewind, who read the passage that had been pointed out by a black fingernail.

He stopped reading, and stared at the Librarian.

It was uplifting. Oh, it was uplifting. Rincewind hadn't read anything like it. But ...

He'd spent the day in this city. There were dog fights and bear pits and that wasn't the worst of it. He'd seen the heads on spikes over the gates. Of course, Ankh-Morpork had been bad, but Ankh-Morpork had thousands of years of experience of being a big city and had become, well, *sophisticated* in its sins. This place was half farmyard.

The man who wrote this woke up every morning in a city that burned people alive and had *still* written this.

'—what a piece of work is a man ... how noble in reason ...

how infinite in faculty ... in form, in moving, how express and admirable ...'

The Librarian was almost sobbing with laughter.

'Nothing to laugh at, it's a perfectly valid point of view,' said Rincewind. He shuffled the pages.

'Who wrote this?' he said.

'According to the flows of L-space, he is widely regarded as one of the greatest playwrights who ever lived,' said Hex, from the shelf.

'What was his name?'

'His own spelling is inconsistent,' said Hex, 'but the consensus is that his name was William Shakespeare.'

'Does he exist on this world?'

'Yes. In one of the many alternate histories.'

'So not actually *here*, then?'

'No. The leading playwright in this city is Arthur J. Nightingale.'

'Is he any good?'

'He is the best they have. Objectively, he is dreadful. His play *King Rufus III* is widely considered the worst play ever written.'

'Oh.'

'Rincewind!' bellowed the Archchancellor.

The wizards were gathering in the circle. They had tied horseshoes and bits of iron to their staffs and had the look of high-order men prepared to kick low-order ass. Rincewind tucked the pages in his robe, picked up Hex and hurried over.

'I'll just—' he began.

'You're coming, too. No arguing. And the Luggage,' snapped Ridcully.

'But—'

'Otherwise we might have a talk about seven buckets of coal,' the Archchancellor went on.

He *knew* about the buckets. Rincewind swallowed.

'Leave Hex behind with the Librarian, will you?' said Ponder. 'He can keep an eye on Dr Dee.'

'Isn't Hex coming?' said Rincewind, alarmed at the prospect of losing the only entity at UU that seemed to have a grasp on things.

'There will be no suitable avatars,' said Hex.

'He means no magic mirrors, no crystal balls,' said Ponder. 'Nothing that people *expect* to be magical. No people at all, where we're going. Put Hex down. We'll be back instantly, in any case. Ready, Hex?'

For a moment the circle glowed, and the wizards vanished.

Dr Dee turned to the Librarian.

'It works!' he said. 'The Great Seal works! Now I can—'

He vanished. And the floor vanished. And the house vanished. And the city vanished. And the Librarian landed in the swamp.

EIGHT

PLANET OF THE APES

 'WHAT A PIECE OF WORK is a man! How noble in reason! How infinite in faculty! In form, in moving, how express and admirable! In action how like an angel! In apprehension how like a god!'

But you wouldn't want to watch him eat, close up …

William Shakespeare was another key figure in the transition from medieval mysticism to post-Renaissance rationalism. We were going to mention him, but we had to wait for him to turn up in Roundworld.

Shakespeare's plays are a cornerstone of our present Western civilisation.* They led us from a confrontation between aristocratic barbarism and tradition-bound tribalism into real civilisation as we know it. And yet … he seems to be a contradiction: uplifting sentiments in a barbarous age. That's because he was standing at a pivotal point in history. The elves have been seeking something that will become human, and will interfere with Roundworld to make sure they get it. Humans are superstitious. But the human condition can also create a Shakespeare. Though not in this version of history.

The elves aren't the only Discworld inhabitants that have interfered with Roundworld: the wizards have tried some 'uplift' of their own, in the sense of David Brin, and using the techniques of Arthur C. Clarke. Near the end of *The Science of Discworld,* the apes of Roundworld are

---

* On his first visit to England in 1930, Mahatma Gandhi was asked 'What do you think of modern civilisation?' He is said to have replied 'That would be a good idea.'

sitting in their cave, watching a manifestation from another dimension, an enigmatic black rectangular slab … The Dean of Unseen University taps on it with his pointer, to attract attention, and chalks the letters R-O-C-K. 'Rock. Can anyone tell me what you do with it?' But all the apes are interested in is S-E-X.

The next time the wizards look at Roundworld, the space elevator is collapsing. The planet's inhabitants are heading out into the universe on vast ships made from the cores of comets.

Something very dramatic has happened between the apes and the space elevator. What was it? The wizards have no idea. They doubt very much it could have had much to do with those apes, who were very much The Wrong Stuff.

In the first volume of *The Science of Discworld*, we explored no further. We left a gap. It was a tiny part of the historical record on the geological timescales that governed everything up to the ape, but rather a big gap in terms of changes to the planet. But now even the wizards are aware that the apes, unpromising material as they may have been, did in fact evolve into the creatures that built the space elevator and fled from a very dangerous planet in search of, as Rincewind would put it, a place where you are not hit on the head with rocks on a regular basis. And, apparently, a key step in their evolution was elvish interference.

How did it actually happen on Roundworld? Here, the whole process took a mere five million years. One hundred thousand Grandfathers* ago, we and the chimpanzees shared a distant ancestor. The chimpanzeelike ancestor of Man was also the Manlike ancestor of the chimpanzee. To us, it would have looked astonishingly like a chimpanzee – but to a chimpanzee, it would have looked astonishingly like a human.

DNA analysis shows, beyond any shadow of reasonable doubt, that our closest living relatives are chimpanzees: the ordinary ('robust')

---

* A time measurement we developed in *The Science of Discworld* as a 'human' way of measuring large amounts of time. It's 50 years, a 'typical' age gap between grandparent and grandchild. Most of the really interesting bits of human development have taken place in the last 150 Grandfathers. Remember – objects in the rear view mirror are closer than they appear.

chimpanzee *Pan troglodytes* and the more slender ('gracile') bonobo *Pan paniscus*, often politically incorrectly called the pygmy chimpanzee. Our genomes have 98 per cent in common with both, leading Jared Diamond to refer to humans as 'the third chimpanzee' in a book of the same title.

The same DNA evidence indicates that we and today's chimpanzees parted company, specieswise, those five million years (100,000 Grandfathers) ago. That figure is debatable, but it can't be very far wrong. The gorillas split off a little earlier. The earliest fossils of our 'hominid' ancestors are found in Africa, but there are numerous later fossilised hominids from other parts of the world such as China and Java. The oldest known are two species of *Australopithecus*, each about 4–4.5 million years old. The Australopithecines had a good run: they hung around until about 1–1.5 million years ago, at which point they gave way to genus *Homo*: *Homo rudolfensis, Homo habilis, Homo erectus, Homo ergaster, Homo heidelbergensis, Homo neanderthalensis*, and finally us, *Homo sapiens*. And somehow another Australopithecine inserted itself into the middle of those *Homo*s. In fact the more hominid fossils we find, the more complicated our conjectured ancestry becomes, and it now looks as if many different hominid species coexisted on the plains of Africa for most of the past five million years.

Today's chimpanzees are quite bright, probably a lot brighter than the apes that the Dean tried to teach spelling to. Some remarkable experiments have shown that chimps can understand a simple version of language, presented to them as symbolic shapes. They can even form simple concepts and make abstract associations, all within a linguistic frame. They can't build a space elevator, and they never will unless they evolve considerably and avoid being killed for 'bush meat'.

We can't build one either, but it might take no more than a couple of hundred years before the things are sprouting all along the equator. All you need is a material with enough tensile strength, perhaps some composite involving carbon nanotubes. Then you dangle cables from geostationary satellites, hang elevator compartments from them, equip them with suitable space elevator music ... after which, leaving the planet becomes entirely straightforward. The energy cost, hence the marginal financial cost, is near enough zero, because for everything that needs to go up, something else needs to come down. It could be moon rock, or platinum mined in the asteroid belt, or the astronaut that the person going up is due to replace on duty. The capital cost of such a project is enormous, though, which is why we're not in any great hurry right now.

The big scientific problem in this connection is: how can evolution get so quickly from an ape that can't compete mentally with a chimpanzee to a godlike being that can write poetry as good as Shakespeare's, and has advanced so rapidly from that point that it will surely soon erect (drop) a space elevator? 100,000 Grandfathers hardly seems long enough, given that it took about 50 million Grandfathers* to get from a bacterium to the first chimpanzee.

Something that dramatic needed a new trick. That trick was the invention of culture. Culture allowed any individual ape to make use of the ideas and discoveries of thousands of other apes. It let the ape collective acquire knowledge cumulatively, so that it didn't all get lost when its owner died. In *Figments of Reality* we coined the term 'extelligence' for this suite of tricks, and the word is beginning to become common currency. Extelligence is like our own personal intelligence, but it lives outside us. Intelligence has limits; extelligence is infinitely expandable. Extelligence lets us pull ourselves upwards, as a group, by our own mental bootstraps.

The contradiction between Shakespeare's noble sentiments and the heads-on-spikes culture in which he lived is a consequence of his position as a very intelligent intelligence in a not-very-extelligent extelligence. Many individuals possessed the nobility to qualify for Shakespeare's praise, but their as yet rudimentary extelligence had not yet transmitted that nobility into the general culture. The culture was, or claimed to be, noble in principle – kings taking their authority from God Himself – but it was a barbarian style of nobility. And it was welded to a barbarian cruelty, the kings' means of self-preservation.

* Most of them being Grandfather bacteria, you appreciate. That's the trouble with metaphors.

There may be many ways to make intelligent creatures, and many more ways to knit them together into an extelligent culture. The crab civilisation in *The Science of Discworld* was doing fine until its Great Leap Sideways was clobbered by an inbound comet. We made that one up, but who knows what might have happened a hundred million years ago? All we know for sure – or for a given value of 'sure', since even now a lot of our knowledge is guesswork – is that some things like apes turned into us. It takes a special kind of arrogance and blindness to extrapolate that story to the rest of the universe without wondering about alternatives.

An important ingredient in *our* story was brains. Weight for weight, humans have far bigger brains than any other animal on the planet. The average human brain has a volume of about 1,350 cubic centimetres, which is roughly three times as great as the brain of apes with the same size body as ours. Whale brains are bigger than ours, but whales are even bigger, so the amount of whale per brain cell is greater than the amount of human per brain cell. When it comes to brains, quantity is less important than quality, of course. But a brain capable of really complicated things like carbon nanotube engineering and fixing dishwashers has to be fairly large, because the abilities of small brains are limited by lack of room to do anything interesting.

We'll see shortly that brains alone are not enough. Nonetheless, without brains, or adequate substitutes, you don't get very far.

There are two main theories of human origins. One is rather dull and probably correct; the other is exciting and most likely wrong. Nevertheless, the second one has quite a lot going for it and is a better story, so let's take a look at them both.

The dull, conventional theory is that we evolved on the savannahs. Roving groups of early apes trailed through the long grass, picking up whatever food – seeds, lizards, insects – they could find, much like today's baboons.* And as they did so, lions and leopards prowled through the long grass looking for monkeys. Those monkeys or apes that were better at spotting the telltale flicker of a big cat's tail, and

---

* Though they're monkeys, not apes.

finding a tree rather quickly, survived to have babies; those that performed poorly at such tasks did not. The babies inherited those survival skills, and passed them on to *their* babies.

What these tasks need is computational power. Spotting a tail and finding a tree are pattern-recognition problems. Your brain needs to pick out the tail-shape from a background of similarly buff-coloured rocks and mud; it has to choose a tree that is tall enough, and climbable enough, without being *too* climbable, and it has to be able to do it fast. A capacious brain with a big memory (of past occasions when something hairy poked out from behind a rock, and of locations for climbable trees) can pick up the visual traces of a lion much more effectively than a small brain can. A brain whose nerve cells transmit signals to each other more quickly can analyse incoming sensory data and conclude 'lion' a lot faster than a slower brain can. So there was evolutionary pressure on the early apes and monkeys to develop bigger and faster brains. There was also evolutionary pressure on the lions to conceal themselves more effectively, so that those bigger and faster ape and monkey brains still didn't notice anything suspicious. So a predator–prey 'arms race' developed, a positive feedback loop that made both lions and apes far more effective in their ecological roles.

That is the conventional story of human evolution. But there is another story, less orthodox, with two main sources.

Human beings are very weird apes, indeed very weird animals altogether. They have extremely short fur, mostly just a downy covering. They walk upright on two legs. They have a layer of fat, all the year round. They mate face-to-face (often). They have exceptionally good breath-control; good enough to be able to speak. They weep and they sweat. They adore water, and can swim long distances. A newborn baby, dropped in a pool, can keep itself afloat: the ability to swim is instinctive. All these peculiarities led Elaine Morgan to write *The Aquatic Ape* in 1982. There she suggested a radical theory: that humans evolved not on the savannahs, surrounded by fierce predators, but on the beach. That explains the swimming, the upright stance (it's easier to evolve a two-legged gait if you are buoyed up by seawater), and the lack of hair (which causes problems when you swim, providing an evolutionary reason for it to disappear). In fact it can be argued that it explains all of the peculiarities of humans that we've just listed. The original scientific underpinnings of this theory were developed by Alister Hardy.

In their 1991 *The Driving Force* Michael Crawford and David Marsh took the story one stage further, by added one extra ingredient. Literally. The most important thing that the beach provides is seafood. And the most important thing that seafood provides is 'essential fatty acids', which are a crucial ingredient in brains. In fact, nearly two-thirds of the human brain is made from them. Fatty acids are good for making membranes, and brains use electrical signals in membranes to compute. Myelin, in a membrane sheath surrounding nerve cells, speeds up the transmission of signals in the human nervous system by a factor of five or so. It takes a lot of essential fatty acids, then, to make a big, fast human brain, so it must have taken almost as much to make the brain of our distant ape ancestor. Oddly, though, our bodies cannot make those special fatty acids from simpler chemicals, like we make most complicated biochemicals that we need. We have to get the fatty acids, ready-made, from our food; that's why the word 'essential' is used to describe them. Even more oddly, there are few essential fatty acids out on the savannahs. They would exist only in living creatures, of course, but even there, they are fairly rare. The richest source of essential fatty acids is seafood.

Perhaps all this explains why we want to spend so much time on the beach. But whatever the explanation, the ability to make big brains was one key step in our evolution away from our hairy, quadrupedal, 100,000-fold great grandfather.

Big brains, however, are not enough. What really matters is what you do with them. And what we managed to do was to play off one brain against another, so that over the millennia they got better and better at competing and communicating.

Ape brains competing with lion brains leads to an arms race that improves both, but the arms race is fairly slow, because both brains are being used for very limited purposes as far as the competition goes. Ape brains competing with other ape brains gives the whole brain a workout, all the time, so the rate of evolution is likely to be much higher.

For every species, the main competition comes from other creatures of the *same* species. This is reasonable; they're the ones that want exactly the same resources that you do. This opens the door to elvish interference, in our Discworld metaphor. The nasty side of human nature, which in extremes leads to evil, is inescapably bound up with the nice side. One very direct way to compete with your neighbour is to bash him on the head, hard.

However, there are more subtle ways to gain evolutionary advancement, as we will see later. The elvish approach is crude, and ultimately self-defeating, for a sufficiently extelligent species.

The possession of brains opens up new non-genetic ways to pass characteristics on to your children. You can give them a good start in life by moulding how their brains react to the outside world. The generic term for this kind of non-genetic transfer between the generations is *privilege*. There are numerous instances of privilege in the animal kingdom. When a mother blackbird provides yolk in her egg for the baby blackbird to feed from, that's privilege. When a cow provides milk for her calf, that's even more privilege. When a mother tarantula wasp provides a paralysed, living spider for her grubs to grow in, that's privilege.

Humans have taken privilege to a qualitatively new level. Human parents invest an astonishing amount of time and effort in their children, and spend decades – entire lifetimes, in many ways – looking after them. In conjunction with big brains, slowly getting bigger as each generation passes, privilege leads to two new tricks, learning and teaching. Those tricks feed off each other, and both require the best brain you can acquire.*

Genes are involved in building brains, and genes can perhaps predispose individuals to be unusually good at learning or teaching. However, both of these educational processes involve far more than mere genes: they take place within a culture. The child does not just learn from its parents. It learns from its grandparents, from its siblings, from its aunts and uncles, from the whole troupe or tribe. It learns, as

---

* It helps considerably to steal privilege from other species; for instance, all that food material in plant seeds, tubers and bulbs.

all parents discover, to their dismay, from undesirable sources as well as authorised ones. Teaching is the attempt to transmit ideas from the adult brain to that of the child; learning is the child's attempt to insert those ideas into its brain. The system is imperfect, with a lot of garbled messages along the way, but despite its faults it is *much* faster than genetic evolution. That's because brains, networks of nerve cells, can adapt much more rapidly than genes can.

The faults, oddly enough, probably accelerate the process, because they are a source of creativity and innovation. An accidental misunderstanding may sometimes lead to an improvement.* In this respect, cultural evolution is just like genetic evolution: it is only because the DNA copying system makes mistakes that organisms can change.

Culture didn't arise in a vacuum: it had many precursors. One crucial step towards the development of culture was the invention of the nest. Before nests came into being, any experimentation by the young either worked, or led to a quick death. Within the protection of the nest, however, young animals can try things out, make mistakes and profit from them; for example, by learning not to do the same thing again. Outside the nest, they never get a chance to try a second time. In this manner nests led to another development, the role of play in educating the young animal. Mother cats bring half-dead mice for their kittens to practise hunting on. Mother birds of prey do the same for their offspring. Polar bear cubs slide down snow-slopes and look cute. Play is good fun, and the kids enjoy it; at the same time, it equips them for their adult roles.

Social animals, ones that gather in groups and operate *as* groups, are a fertile breeding-ground for privilege and for education. And with appropriate communication, groups of animals can achieve things that no individual can manage. A good example is dogs, which evolved the ability to hunt in packs. When such tricks are being played, it is important to have some recognition signal that lets the pack distinguish its own members from outsiders, otherwise the pack can do all the work and then an outsider can steal the food. Each dog pack has its own call-sign, a special howl that only insiders know. The more

* It happens all the time on Discworld!

elaborate your brain, the more elaborate the communication from brain to brain can be, and the more effectively education works.

Communication helps with the organisation of group behaviour, and it opens up survival techniques that are more subtle than bashing others on the head. Within the group, cooperation becomes a far more viable option. Today's great apes generally work as small groups, and it seems likely that their ancestors did the same. When humans split off from the chimpanzee lineage, those groups became what we now call tribes.

Competition between tribes was intense, and even today some jungle tribes in South America and New Guinea think nothing of killing anyone they meet who comes from a different tribe. This is a reversion to the 'bash on the head' option, but now one group cooperates to bash the other group's members on the head. Or, usually, *one* such member at a time. Less than a century ago, most such tribes did the same (one of the stories we've told ourselves throughout our tribal history is that we are The People, The True Human Beings – which means that everyone else isn't).

Chimpanzees have been observed killing other chimpanzees, and they regularly hunt smaller monkeys for meat. That isn't cannibalism. The food is a different species. Most humans cheerfully consume other mammals, even quite intelligent ones like pigs.*

Just as dog-packs need an agreed recognition signal to identify their members, so each tribe needs to establish a distinct identity. The possession of big brains makes it possible to do this by means of elaborate, shared rituals.

Ritual is by no means confined to humans: many species of birds, for instance, have special mating dances, or engage in strange devices to attract the female's attention, like the decorative collections of berries and pebbles assembled by the male bower-bird. But humans, with their highly developed brains, have turned ritual into a way of life. Every tribe, and nowadays every culture, has developed a Make-a-Human kit whose object is to bring up the next generation to adopt the tribal or cultural norms and pass them on to their own children.

---

* But we eat sheep, too.

It doesn't always work, especially nowadays when the world has shrunk and cultures clash across non-geographical boundaries – Iranian teenagers accessing the Internet, for example – but it still works surprisingly well. Corporations have taken up the same idea, with 'corporate bonding' sessions. This is what the wizards were up to with their paintballs. Studies have shown that sessions of this kind have no useful effect, but businesses still waste billions on them every year. The second most probable reason is that such sessions are fun anyway. The first most probable is that everyone likes an opportunity to shoot Mr Davis in Human Resources. And one important reason is that it *sounds* as though it ought to work; our culture is full of stories where such things do.

An important part of the Make-a-Human kit is the Story. We tell our children stories, and through those stories they learn what it is like to be a member of our tribe or our culture. They learn from the story of Winnie the Pooh getting stuck in Rabbit's hole that greed can lead to constraints on food. From the Three Little Pigs (a civilising story, not a tribal one) they learn that if you watch your enemy for repetitive patterns, you can outwit him. We use stories to build our brains, and then we use the brains to tell ourselves, and each other, stories.

As time passes, those tribal stories acquire their own status, and people cease to question them because they are traditional tribal stories. They acquire a veneer of – well, the elves would call it 'glamour'. They seem wonderful, despite numerous obvious faults, and most people do not question them. On Discworld, precisely this process occurred with stories and folk-memories about elves, as we can illustrate with three quotations from *Lords and Ladies*. In the first, the god of all small furry prey, Herne the Hunted, has just come to the terrified realisation that '*They're all coming back!*'. Jason Ogg, who is a blacksmith, the eldest son of the witch Nanny Ogg, and not very bright, asks her who *They* are:

> 'The Lords and Ladies,' she said.
> 'Who're they?'

Nanny looked around. But, after all, this was a forge ... It wasn't just a place of iron, it was a place where iron died and was reborn. If you couldn't speak the words here, you couldn't speak 'em anywhere.

Even so, she'd rather not.

'*You* know,' she said. 'The Fair Folk. The Gentry. The Shining Ones. The Star People. *You* know.'

'What?'

Nanny put her hand on the anvil, just in case, and said the word.

Jason's frown very gently cleared, at about the same speed as a sunrise.

'Them?' he said. 'But aren't they nice and—'

'See?' said Nanny. 'I *told* you you'd get it wrong!'

You said: The Shining Ones. You said: The Fair Folk. And you spat, and touched iron. But generations later, you forgot about the spitting and the iron, and you forgot why you used those names for them, and you remembered only that they were beautiful ... We're stupid, and the memory plays tricks, and we remember the elves for their beauty and the way they move, and we forget what they *were*. We're like mice saying, 'Say what you like, cats have got real *style*.'

Elves are wonderful. They provoke wonder.

Elves are marvellous. They cause marvels.

Elves are fantastic. They create fantasies.

Elves are glamorous. They project glamour.

Elves are enchanting. They weave enchantment.

Elves are terrific. They beget terror.

The thing about words is that meanings can twist just like a snake, and if you want to find snakes look for them behind words that have changed their meaning.

No-one ever said elves are *nice*.

Elves are *bad*.

For most purposes (though, admittedly, not when dealing with elves) it doesn't greatly matter if the traditional tales make no real sense. Father Christmas and the Tooth Fairy make no immediate sense (on

Roundworld, but see *Hogfather* for their Discworld significance). Mind you, it's clear why children are happy to believe in such generosity. The most important role of the tribal Make-a-Human kit is to provide the tribe with its own collective identity, making it possible for it to act as a unit. Tradition is good for such purposes; sense is optional. All religions are strong on tradition, but many are weak on sense, at least if you take their stories literally. Nevertheless, religion is absolutely central to most cultures' Make-a-Human kit.

The growth of human civilisation is a story of the assembly of ever-larger units, knitted together by some version of that Make-a-Human kit. At first, children were taught what they must do to be accepted as members of the family group. Then they were taught what they must do to be accepted as members of the tribe. (Believing apparently ridiculous things was a very effective test: the naïve outsider would all too readily betray a lack of belief, or would simply have no idea what the appropriate belief was. Is it permitted to pluck a chicken before dark on Wednesday? The tribe knew, the outsider did not, and since any reasonable person would guess 'yes', the tribal priesthood could go a long way by making the accepted answer 'no'.) After that, the same kind of thing happened for the local baron's serfs, for the village, the town, the city and the nation. We spread the net of True Human Beings.

Once units of any size have acquired their own identity, they can *function* as units, and in particular they can combine forces to make a bigger unit. The resulting structure is hierarchical: the chains of command reflect the breakdown into sub-units and sub-sub-units. Individual people, or individual sub-units, can be expelled from the hierarchy, or otherwise punished, if they stray outside accepted (or enforced) cultural norms. This is a very effective way for a small group of people (barbarian) to maintain control over a much larger group (tribal). It works, and because of that we still labour under its restrictions, many of which are undesirable. We have invented techniques like democracy to try to mitigate the undesirable effects, but these techniques bring new problems. A dictatorship can generally take action more rapidly than a democracy, for example. It's harder to argue.

The path from ape to human is not just one of evolutionary pressures producing more and more effective brains; not just a tale of the evolution of intelligence. Without intelligence, we could never have got started on that path, but intelligence alone was not enough. We had to find a way to share our intelligence with others, and to store useful ideas and tricks for the benefit of the whole group, or at least, those in a position to make use of it. That's where extelligence comes into play. Extelligence is what really gave those apes the springboard that would launch them into sentience, civilisation, technology, and all the other things that make humans unique on this planet. Extelligence amplifies the individual's ability to do good – or evil. It even creates new forms of good and evil, such as, respectively, cooperation and war.

Extelligence operates by putting ever more sophisticated stories into the Make-a-Human-Being kit. It pulled us up by our own bootstraps: we could climb from tribal to barbarian to civilised.

Shakespeare shows us doing it. His period was *not* a rebirth of Hellenistic Greece or Imperial Rome. Instead, it was the culmination of the barbarian ideas of conquest, honour and aristocracy, codified in the principles of chivalry, meeting its match in the written principles of a tribal peasantry, and disseminated by printing. This kind of sociological confrontation produced many events in which the two cultures meet head on.

This was exemplified by the Warwickshire enclosure uprisings. In Warwickshire, the aristocracy carved up land into small parcels, and the peasantry got very upset because the aristocracy didn't give any heed to what kind of land was in each parcel. All the aristocrats knew about peasant farming was a simplistic calculation: this much land will suffice for that many peasants. The peasants knew what was actually involved in growing food, so that the only thing you could do with a small piece of woodland, for instance, was to chop down all the trees to make room to grow some food.

Today's bean-counting managerial style in many businesses, and all British public services, is exactly the same. This kind of confrontation between the barbarian attitudes of the nobility and the tribal ones of the peasantry is precisely portrayed in many of Shakespeare's plays, as an illustration of low-life, with its folk wisdom as comic relief and pathos, set against the lofty ideals of the ruling classes – leading so often to tragedy.

But also to high comedy. Think of Theseus, Duke of Athens, on the one hand, and Bottom on the other, in *A Midsummer Night's Dream*.

NINE

THE ELVISH
QUEEN

IN THE HEAT OF THE NIGHT, magic moved on silent feet.

One horizon was red with the setting sun. This world went around a central star. The elves did not know this. If they had done, it would not have bothered them. They never bothered with detail of that kind. The universe had given rise to life in many strange places, but the elves were not interested in that, either.

This world had created lots of life, too. None of it had ever had what the elves considered to be potential. But this time ...

It had iron, too. The elves hated iron. But this time, the rewards were worth the risk. This time ...

One of them signalled. The prey was close at hand. And now they saw it, clustered in the trees around a clearing, dark blobs against the sunset.

The elves assembled. And then, at a pitch so strange that it entered the brain without the need to use the ears, they began to sing.

'Charge!' cried Archchancellor Ridcully.

The wizards, all bar Rincewind, charged. He peered around from behind a tree.

The elf song, a creative dissonance of tones that went straight into the back of the brain, ceased abruptly.

Thin figures spun around. Almond eyes glowed in triangular faces.

People who knew the wizards only as the world's most avid diners would have been quite surprised at their turn of speed. Besides, while it may take a little while for a wizard to reach maximum acceleration, he's then very hard to stop. And he carries such a cargo of aggression; the stratagems of the Uncommon Room at UU are guaranteed to give any wizard a maximum load of virulence just itching for a target.

The Dean hit first, striking an elf a blow with his staff. A horseshoe had been wired to the end. The elf screamed and twisted back, clutching at its shoulder.

There were many elves but they hadn't been expecting an attack. And iron was so powerful. A handful of flung nails had the effect of buckshot. Some tried to fight back, but the dread of iron was too strong.

The prudent and the survivors took to their skinny heels, while the dead evaporated.

The attack took less than thirty seconds. Rincewind watched it from behind his tree. He was not being cowardly, he reasoned. This was a job for specialists, and could safely be left to the senior wizards. If, later on, there was a problem involving slood dynamics or fretwork, or someone needed to misunderstand some magic, he would be happy to step forward.

There was a rustling behind him.

Something was there. What it was *changed* as he turned and stared.

The first talent of the elves was their singing. It could turn other creatures into potential slaves. The second talent was their ability to change not their shape but how their shape was perceived. For a moment Rincewind caught sight of a slim, spare figure glaring at him and then, in one blurred moment, it became a woman. A queen, in a red dress and a rage.

'Wizards?' she said. 'Here? Why? How? Tell me!'

A gold crown glittered in her dark hair and murder gleamed in her eyes as she advanced on Rincewind, who backed up against his tree.

'This is *not* your world!' the elf queen hissed.

'You'd be amazed,' said Rincewind. 'Now!'

The queen's brow wrinkled. 'Now?' she repeated.

'Yes, I said *now*,' Rincewind said, grinning desperately. '*Now* was the word I said, in fact. *Now!*'

For a moment the queen looked puzzled. And then she somersaulted backwards in a high arc, just as the Luggage's lid snapped shut where she had been. She landed behind it, hissed at Rincewind, and vanished into the night.

Rincewind glared at the box. 'Why did you wait? Did I tell you to wait?' he demanded. 'You just like to stand right behind people and wait for them to find out, right?'

He looked around. There was no sign of any more elves. In the middle distance the Dean, having run out of enemies, was attacking a tree.

Then Rincewind looked up. Along the branches, clinging to one another and staring down at him in wide-eyed amazement, were dozens of what looked, in the moonlight, like rather small and worried monkeys.

'Good evening!' he said. 'Don't worry about us, we're just passing through ...'

'Now this is where it all gets complicated,' said a voice behind him. It was a familiar one, being his own. 'I've only got a few seconds before the loop closes, so here's what you have to do. When you go back to Dee's time ... hold your breath.'

'Are you *me*?' said Rincewind, peering into the gloom.

'Yes. And I'm telling you to hold your breath. Would I lie to me?'

There was an inrush of air as the other Rincewind disappeared and, down in the clearing, Ridcully bellowed Rincewind's name.

Rincewind stopped looking around and hurried down to the other wizards, who were looking immensely pleased with themselves.

'Ah, Rincewind, I thought you wouldn't want to be left behind,' said the Archchancellor, grinning nastily. 'Got any, did you?'

'The queen, in fact,' said Rincewind.

'Really? I'm impressed!'

'But she – *it* got away.'

'They've all gone,' said Ponder. 'I saw a blue flash on that hill up there. They've gone back to their world.'

'D'you think they'll come back?' said Ridcully.

'It doesn't matter if they do, sir. Hex will spot them and we can always get there in time.'

Ridcully cracked his knuckles. 'Good. Capital exercise. Much better than magicking paint at one another. Builds grit and team interdependency. Someone go and stop the Dean attacking that rock, will you? He does rather get carried away.'

A faint white ring appeared on the grass, wide enough to hold the wizards.

'Ah, the ride back,' said the Archchancellor, as the excited Dean was hustled towards the rest of the group. 'Time to—'

The wizards were suddenly in empty air. They fell. All but one of them were not holding their breath before they hit the river.

Wizards do, however, have good floating capabilities and a tendency to bob up and down. And the river was in any case rather like a slowly moving swamp. Floating logs and mud banks choked it. Here and there, mud banks had become sufficiently established to sprout a crop of trees. By degrees, and with much arguing about where dry land actually began – it was not very obvious – they splashed their way to the shore. The sun was hot overhead, and clouds of mosquitoes shimmered among the trees.

'Hex has brought us back to the wrong time,' said Ridcully, wringing out his robe.

'I don't think he'd do that, Archchancellor,' said Ponder, meekly.

'The wrong place, then. This is *not* a city, in case you hadn't noticed.'

Ponder looked around in bewilderment. The landscape around them was not exactly land and not exactly river. Ducks were quacking, somewhere. There were blue hills in the distance.

'On the upside,' said Rincewind, extracting frogs from a pocket, 'everything smells better.'

'This is a *swamp*, Rincewind.'

'So?'

'And I can see smoke,' said Ridcully.

There was a thin grey column in the middle distance.

Reaching it took a lot longer than the mere distance suggested. Land and water were contesting every step of the way. But, eventually, and with only one sprain and a number of bites, the wizards reached some thick bushes and peered into the clearing beyond.

There were some houses, but that was stretching the term. They were little more than piles of branches with reed roofs.

'They could be savages,' said the Lecturer in Recent Runes.

'Or perhaps someone sent them all out into the country to forge a dynamic team spirit,' said the Dean, who had been badly bitten.

'Savages would be too much to hope for,' said Rincewind, watching the huts carefully.

'You *want* to find savages?' said Ridcully.

Rincewind sighed. 'I *am* the Professor of Cruel and Unusual Geography, sir. In an unknown situation, *always* hope for savages. They tend to be quite polite and hospitable provided you don't make any sudden moves or eat the wrong sort of animal.'

'Wrong sort of animal?' said Ridcully.

'Taboo, sir. They tend to be related. Or something.'

'That sounds rather ... sophisticated,' said Ponder suspiciously.

'Savages often are,' said Rincewind. 'It's the civilised people that give you trouble. They always want to drag you off somewhere and ask you unsophisticated questions. Edged weapons are often involved. Trust me on this. But these aren't savages, sir.'

'How can you tell?'

'Savages build better huts,' said Rincewind firmly. 'These are edge people.'

'I've never heard of edge people!' said Ridcully.

'I made it up,' said Rincewind. 'I run into them occasionally. People that live right on the edge, sir. Out on rocks. In the worst kind of desert. No tribe or clan. That takes too much effort. Of course, so does beating up strangers, so they're the best kind of people to meet.'

Ridcully looked around at the swamp. 'But there's waterfowl everywhere,' he said. 'Birds. Eggs. Lots of fish, I'll be bound. Beavers. Animals that come down to drink. I could eat myself greasy to the eyebrows here. This is *good* country.'

'Hold on, one of them's coming out,' said the Lecturer in Recent Runes.

A stooping figure had emerged from a hut. It straightened up, and stared around. Huge nostrils flared.

'Oh my, look what just fell out of the ugly tree,' said the Dean. 'Is it a troll?'

'He's certainly a bit rugged,' said Ridcully. 'And why is he wearing boards?'

'I think he's just not very good at tanning hides,' said Rincewind.

The enormous shaggy head turned towards the wizards. The nostrils flared again.

'He smelled us,' said Rincewind, and started to turn. A hand grabbed the back of his robe.

'This is not a good time to run away, *Professor*,' said Ridcully, lifting him off the ground in one hand. 'We know you're good at languages. You get on with people. You have been chosen to be our ambassador. Do not scream.'

'Besides, the thing looks like cruel and unusual geography,' said the Dean, as Rincewind was thrust out of the bushes.

The big man watched him, but made no attempt to attack.

'Go *on*!' hissed the bushes. 'We need to find out when we are!'

'Oh, right,' said Rincewind, eyeing the giant cautiously. 'And he's going to tell me, is he? He's got a calendar, has he?'

He advanced carefully, hands up to show that he didn't have a weapon. Rincewind was a great believer in not being armed. It made you a target.

The man had obviously seen him. But he didn't seem very excited about it. He watched Rincewind as someone might watch a passing cloud.

'Er ... hello,' said Rincewind, stopping just out of range. 'Me big fella Professor of Cruel and Unusual Geography belong Unseen University, you ... oh dear, you haven't even discovered washing, have you? Either that or it's the clothes belong you. Still, no obvious weapons. Er ...'

The man took a few steps forward and tugged the hat off Rincewind's head in one quick movement.

'Hey—!'

What was visible of the big face broke into a grin. The man turned the hat this way and that. Sunlight sparkled off the word 'Wizzard' in cheap sequins.

'Oh, I see,' said Rincewind. 'Pretty glitter. Well, that's a start ...'

TEN

BLIND MAN
WITH LANTERN

 THE WIZARDS ARE NOW BEGINNING to understand that, while you *can* eliminate evil by eliminating extelligence, the result can be about as interesting as watching daytime television. Their plan to stop the elves interfering with human evolution has worked, but they don't like the result. It is bland and unintelligent. It has no spark of creativity.

How did human creativity arise? By now you won't be surprised to learn that it came from stories. Let's take a closer look at the current scientific view of human evolution, and fill in that gap between R-O-C-K and the space elevator.

An elf, observing Earth's landmasses 25 million years ago, would have seen vast areas of forest. From the hills of northern India to Tibet and China, and down into Africa, these forests held a great variety of small apes, ranging from about half the size of chimpanzees to the size of gorillas. The apes were at home on the ground and in the lower branches of the forest, and they were so common that today we have many fossils of them. In addition, the Old World monkeys were starting to diversify in the upper levels of the forest. Earth was a Monkey Planet.

But also a Snake Planet, a Big Cat Planet, a Nematode Planet, an Alga Planet and a Grass Planet. Not to mention Plankton Planet, Bacterium Planet and Virus Planet. The elf might not have noticed that the African apes had produced several ground-dwelling kinds, not very different from the monkey-derived baboons. And it might also have failed to spot the presence of gibbons in the high branches, alongside the monkeys. These creatures were not particularly remarkable against a background of spectacular large mammals like rhinoceroses, a variety of forest elephants, bears. But we humans are interested in them, because they were our ancestors.

We call them 'woods-apes', dryopithecines. Some, known as *Ramapithecus,* were of lighter build – the jargon is 'gracile'. Others, such as *Sivapithecus*, were big and strong – 'robust'. The lineage of *Sivapithecus* was the one that led to orangutans. These early apes would have been shy, morose creatures like today's wild apes, occasionally playful, but the adults would have been very belligerent and conscious of status within the group.

The forests inhabited by the woods-apes slowly dwindled as the climate cooled and dried, and grasslands – savannah country – took over. There were ice ages, but in the region of the tropics these did not reduce temperatures severely. However, they did change the patterns of rainfall. The monkeys thrived, producing many ground-living kinds like baboons and vervets, and the ape populations got smaller.

By ten million years ago, there were few apes left. There are almost no fossil apes from that period. It seems plausible that, as now and as previously, those apes that did still exist were forest creatures. Some, like today's chimpanzees, gorillas and orangutans, were probably common in a few locations in the forests, but you'd have needed a lot of luck to find them. The observing elf might, even then, have put *all* of these apes on its Endangered List of Earth Mammals. Like very nearly all animal groups that had evolved, the forest apes were soon to be history rather than ecology. The common ancestor of humans and chimpanzees was, then, a not very remarkable ape that probably lived much as the different chimpanzees do now: some in flooded forest like today's bonobos, some in rain-forest, and some in fairly open woodland grading into grasslands. The gorilla lineage separated from the other apes around this time.

At first, the elf would probably not have been very interested as – according to one of the two popular theories of human origins – a new kind of ape began to evolve a more upright stance than those of its relatives, lost its hair, and moved out on to the savannah. Many other mammals did the same; there was a new kind of living to be made on the great grass plains. Giant hyenas, massive wild dogs, lions and cheetahs made a good living from the vast herds of herbivores that lived on the productive savannah grasses; the giant pythons were probably originally savannah animals, too.

The story has been told many times, in many versions. And that's just the point: we understand our ancestry *through* story. We wouldn't be able to work out our ancestry from the fossils that we have discovered unless we'd learned just what clues to look for, especially since few fossil sites have enough evidence left.

The new ancestral plains ape saw the world differently. Judging from the behaviour of today's chimpanzees, especially bonobos, it was a highly intelligent animal. We call their fossils southern apes, australopithecines, and there are hundreds of books that tell stories about them. They may have sojourned by the sea, doing clever things on beaches. Some certainly lived on lake margins. Today's chimpanzees use stones to smash hard nuts open, and sticks to extract ants from nests; the australopithecines also used stones and sticks as tools, rather more so than their cousins the chimpanzees now do. They may have killed small game, as chimpanzees do. They probably used sexual behaviour to hang much of their pleasure on, like today's bonobos, but most likely they were more gender-conscious and male-dominant. Like previous apes, they diverged into gracile and robust lines. The robust ones, called *Anthropithecus boisi*, or even a different genus *Zinjanthropus* ('nutcracker man') and other defamatory names, were vegetarians like today's gorillas, and probably left no descendants in modern times.

This kind of split into gracile and robust forms, by the way, seems to be one of the standard patterns of evolution. Mathematical models suggest that it probably happens when a mixed population of big and small creatures can exploit the environment more effectively than a single population of medium-sized ones, but this idea has to be considered highly speculative until more evidence comes in. The zoological world was recently given a reminder of how common such a split is, and of how little we really know about the creatures of our own planet.

The animal involved could not have been better known, nor more appropriate to Discworld: the elephant.* As every child learns at an early age, there are two kinds of elephant, two distinct species: the African elephant and the Indian elephant.

Not so. There are three species. Zoologists have been arguing for at least a century about what they thought was at most a subspecies of 'the' African elephant *Loxodonta africana*. The typical big, burly African elephant lives on the savannah. The elephants that live in the forest are shy, and difficult to spot: there is just one of them in the Paris zoo, for example. Biologists had assumed that because the forest elephants and the savannah elephants can interbreed at the edges of the forest, they could not be separate species. After all, the standard definition of a species, promoted by the evolutionary biologist Ernst Mayr, is 'able to interbreed'. So they either insisted that there was just one species, or that 'African elephant' had a distinguished subspecies, the forest elephant *Loxodonta africana cyclotis*. On the other hand, zoologists who have had the good fortune to see forest elephants are in no doubt that they look very different from the savannah ones: they are smaller, with straighter, longer tusks, and round ears, not pointed ones. Nicholas Georgiadis, a biologist at the

---

* There's been a very cute discovery about elephants recently, and the only place we can find to put it is this footnote. (This, after all, is what footnotes are for.) It has been known since 1682 that elephants' lungs are unusual, without the 'pleural cavity', a space between the lungs and the chest wall that is filled with fluid, that most mammals have. Instead of fluid, elephants' lungs are surrounded by loose connective tissue. It now looks as if this type of lung exists because it lets elephants go snorkelling, breathing through their trunks. In 2001 the physiologist John West calculated that with a normal pleural cavity, the pressure of the water would burst the tiny blood vessels in the pleural membrane and snorkelling could be fatal. We're now wondering whether the trunk evolved in the ocean as a snorkel. Land vertebrates first evolved from fish that came up on to the seashore. Much later, a variety of mammals went back into the oceans and evolved into several kinds of sea-mammals, the most spectacular modern descendants being whales. We now see that somewhere along the way, some of those water-adapted mammals came back on to the land and turned into elephants. So the elephant is now on its *second* evolutionary journey out of the water and on to the land. It would be nice if it made up its mind.

Mpala Research Centre in Kenya, has said: 'If you see a forest elephant for the first time, you think, "Wow, what is that?"' But because biologists *knew*, on theoretical grounds, that the animals *had* to be all the same species, the observational evidence was rejected as inconclusive.

However, in August 2001 a team of four biologists – Georgiadis, Alfred Roca, Jill Pecon-Slattery and Stephen O'Brien – reported in the journal *Science* their 'Genetic evidence for two species of elephant in Africa'. Their DNA analysis makes it absolutely clear that the African elephant really does come in two distinct forms: the usual robust form, and a separate gracile form. Moreover, the gracile African elephants really are a different species from the robust ones. As different, in fact, as either African species is from the Indian one. So now we have the robust African plains elephant *Loxodonta africana* and the gracile African forest elephant *Loxodonta cyclotis*.

What of the belief that there could be only one species because of the potential for interbreeding? This particular definition of species is taking a hammering at the moment, and deservedly so. The main reason is a growing realisation that even when animals *can* interbreed, they may decide not to.

The story of the Third Elephant is not new: only the names have been changed. Before 1929 every zoologist 'knew' there was only one species of chimpanzee; after 1929, when the bonobos of the inaccessible swamps of Zaire were recognised as a second species,* it became obvious to many zoos that they had possessed two distinct chimpanzee species for years, but not realised it. Exactly the same story is now being played out with elephants.

As we've mentioned, Discworld recently revived interest in its fifth elephant, a story told, you will be surprised to hear, in *The Fifth Elephant*. According to legend, there were originally five elephants standing on Great A'Tuin and supporting the Disc, but one slipped, fell off the turtle, and crashed into a remote region of Discworld:

They *say* that the fifth elephant came screaming and trumpeting

---

* See *The Science of Discworld*, chapter 38.

through the atmosphere in the young world all those years ago and landed hard enough to split continents and raise mountains.

No one actually saw it land, which raised the interesting philosophical question: when millions of tons of angry elephant come spinning through the sky, and there is no one to hear it, does it – philosophically speaking – make a noise?

And if there was no one to see it hit, did it *actually* hit?

There is evidence, in the form of vast deposits of fat and gold (the great elephants that support the world do not have ordinary bones), deep underground in the Schmaltzberg mines. However, there is a more down-to-Disc theory: some catastrophe killed off millions of mammoths, bison and giant shrews, and then covered them over. On Roundworld, there would be a good scientific test to distinguish the two theories: are the deposits of fat shaped like a crash-landed elephant? But there's no point even in looking, on Discworld, because narrative imperative will ensure that they are, even if they were formed by millions of mammoths, bison and giant shrews. Reality has to follow the legend.

Roundworld has so far reached only its third elephant, although Jack hopes that some careful selective breeding might yet bring back a fourth: the pygmy elephant, which lived in Malta and was about the size of a Shetland pony. It would make a marvellous pet – except that, like many diminutive creatures, it would probably be rather bad-tempered. And the very devil to discourage from getting on the settee.

We are a gracile ape (not that you'd notice in some parts of the world, where many of us more closely resemble a robust hippopotamus). About four million years ago one gracile lineage of apes started to get bigger brains and better tools. Against all the rules of taxonomy we call this lineage, our lineage, *Homo*: it really should be *Pan*, because we are the third chimpanzee. We use this name because it is certainly our own lineage, and we prefer to think of ourselves as being enormously different from the apes. In this we could be right: we may indeed share 98 per cent of our genes with chimpanzees, but then, we share 47 per cent with cabbages. Our big difference from the apes is cultural, not genetic. Anyway, within the *Homo* lineage we again find gracile and robust stocks. *Homo habilis* was our gracile tool-making ancestor, but *Homo ergaster* and others went the vegetarian, robust way. If there actually is a yeti or a bigfoot, the best bet is a robust *Homo*. From *Homo habilis*'s success, a larger-brained *Homo* spread out over Africa, into Asia (as Peking Man) and Eastern Europe about 700 million years ago.

We have labelled one variety of these fossils *Homo erectus*. The visiting elf would certainly have noticed this fellow. He had several kinds of tools, and he used fire. He may even have possessed language, of a kind. What we have every reason to suspect that he did, that his ancestors and cousins only occasionally achieved, was to 'understand' his world and change it. Chimpanzees engage in quite a lot of 'if ... then' activities, including 'lying': 'if I pretend not to have seen that banana, I can come back and get it later when that big male won't steal it from me'.

The young of this early hominid grew up in family groups where things were happening that were unlike anything anywhere else on the planet. Sure, there were lots of other mammal nests, packs and troupes, where the young were playing at being adult or just fooling around; nests were safe, and trial-and-error was rarely lethal, so the young could learn safely. But in the human lineage, father was making stone tools, grunting at his women *about* the children, *about* the cave, *about* putting more wood on the fire. There would be favourite gourds *for* banging, perhaps *for* fetching water, spears *for* hunting, lots of stones *for* tools.

Meanwhile, in Africa, another new lineage had arisen about 120,000 years ago, and spread; we call it ancient *Homo sapiens*, and it led to us. Its brain was even bigger, and in caves on the coast of South Africa it – we – had begun to make better tools, and to make primitive paintings on rocks and cave walls. Our population exploded, and we migrated. We reached Australia just over 60,000 years ago, and Europe about 50,000 years ago.

In Europe there had been a moderately robust *Homo*, the Neanderthal *Homo sapiens neanderthalensis*, a subspecies. Some anthropologists consider us to be a sister subspecies, *Homo sapiens sapiens* or, loosely speaking, 'Seriously wise man'. Wow. The Neanderthals' stone tools were well developed for various functions, but these particular hominids seem not to have been progressive. Their culture hardly changed over tens of thousands of years. But they seem to have had some kind of spiritual impulse, for they buried their dead with ceremony – or, at least, with flowers.

Our more gracile ancestors, the Cro-Magnon people, lived at the same time as the last of the Neanderthals, and there are many theories about what happened when the two subspecies interacted. Basically, we survived and the Neanderthals didn't …

Why? Was it because we bashed them on the head more effectively than they bashed us? Did we outbreed them? *In*breed them? Squeeze them out into the 'edge country'? Crush them with superior extelligence? We'll put our own theory forward later in the book.

We don't subscribe to the 'rational' story of human evolution and development, the story that has named us so pretentiously *Homo sapiens sapiens*. Briefly, that story focuses on the nerve cells in our brains, and says that our brains got bigger and bigger until finally we evolved Albert Einstein. They did, and we did, and Albert was indeed pretty bright, but nonetheless the thrust of that story is nonsense, because it doesn't discuss why, or even how, our brains got bigger. It's like describing a cathedral by saying 'You start with a low wall of stones and as time passes you add more stones so that it gets higher and higher'. There's a lot more to a cathedral than that, as its builders would attest.

What actually happened is much more interesting, and you can see it going on all around you today. Let's look at it from the elf's viewpoint. We don't programme our children rationally, as we might set up a computer. Instead, we pour into their minds loads of irrational junk about sly foxes, wise owls, heroes and princes, magicians and genies, gods and demons, and bears that get stuck in rabbit-holes; we frighten them half to death with tales of terror, and they come to enjoy the fear. We beat them (not very much in the last few decades, but for thousands of years before that, for sure). We embed the teaching messages in long sagas, in priestly injunctions, and invented histories full of dramatic lessons; in children's stories that teach them by indirection. Stand near a children's playground, and watch (these days, check with the local police station first, and in any case be sure to wear protective clothing). Peter and Iona Opie did just that, many years ago, and collected children's songs and games, some of them thousands of years old.

Culture passes through the whirlpool that is the child community without needing adults for its transmission: you will all remember Eeny-Meeny-Miny-Mo, or some other counting-out ritual. There is a children's subculture that propagates itself without adult intervention, censorship, or indeed knowledge.

The Opies later collected, and began to explain to adults, the original nursery stories like Cinderella and Rumpelstiltskin. In late medieval times, Cinderella's slipper had been a fur one, not glass. And that was a euphemism, because (at least in the German version) the girls gave the prince their 'fur slipper' to try on … The story came to us through the French, and in that language *verre* can either be 'glass' or 'fur'. The Grimm brothers went for the hygienic alternative, saving parents the danger of embarrassing explanations.

Rumpelstiltskin was an interestingly sexual parable, too, a tale to programme the idea that female masturbation leads to sterility. Remember the tale? The miller's daughter, put in the barn to 'spin straw into gold', virginally *sits on* a little stick that becomes a little man … The dénouement has the little man, when his name is finally identified, jumping in to 'plug' the lady very intimately, and the assembled soldiers can't pull him out. In the modern bowdlerised version, this survives vestigially as the little man pushing his foot through the floor and not being able to pull it out, a total *non sequitur*. So none of those concerned, king, miller or queen, can procreate (the stolen first child has been killed by the soldiers), and it all ends in tears. If you doubt this interpretation, enjoy the indirection: 'What is his name? What is his name?' recurs in the story. What *is* his name? What is a stilt with a rumpled skin? Whoops. The name has an equivalent derivation in many languages, too. (In Discworld, Nanny Ogg claimed to have written a children's story called 'the Little Man Who Grew Too Big', but, then, Mrs Ogg always believed that a *double entendre* can mean only one thing.)

Why do we like stories? Why are their messages embedded so deeply in the human psyche?

Our brains have evolved to understand the world through patterns. These may be visual patterns, such as the tiger's stripes, or aural ones, like the howl of the coyote. Or smells. Or tastes. Or narratives. Stories are little mental models of the world, sequences of ideas strung like beads on a necklace. Each bead leads inexorably to the next bead; we *know* that the second little pig is going to get the chop: the world would not be working properly if it didn't.

We deal not just in patterns, but also in meta-patterns. Patterns of patterns. We watch archer-fish shooting down insects with jets of water, we enjoy the elephant using its nose to acquire doughnuts from zoo visitors (less these days, alas); we delight in the flight of house-martins (there are fewer swallows to enjoy now) and the songs of garden-birds. We admire the weaver birds' nests, the silk moths' cocoons, the cheetahs' speed. All these things are characteristic of the creatures concerned. And what is characteristic of us? Stories. So, by the same token, we enjoy the stories of people. We are the storytelling chimpanzee, and we appreciate the meta-pattern involved in that.

When we became more social, collecting into groups of a hundred or more, probably with agriculture, more stories appeared in our extelligence, to guide us. We had to have rules for behaviour, ways to deal with the infirm and the handicapped, ways to divert violence. In early and present-day tribal societies, everything that is not forbidden is mandatory. Stories point to difficult situations, like the Good Samaritan story in the New Testament; the Prodigal Son, too, is instructive by indirection, like Rumpelstiltskin. To drive that home, here is a tale from the Nigerian Hausa tribe, *Blind Man's Lantern*.

A young man is coming home late from seeing his girlfriend in the next village; it is very dark under a starry sky and the path back to his own village is not easy to follow. He sees a lantern bobbing towards him, but when it gets closer he sees that it is carried by the Blind Man of his own village.

'Hey, Blind Man,' he says. 'You whose darkness is no darker than your noonday! What do *you* carry a lantern for?'

'It is not for my need I carry this lantern,' says the Blind Man. 'It is to keep off you fools with eyes!'

We, as a species, don't only specialise in storytelling. Just as with the other specialities above, our species has a few more oddities. Probably the most odd characteristic that our elvish observer would note is our obsessive regard for children. We not only care for our *own* children, which is entirely to be expected biologically, but for other people's children, too; indeed for other humankind's children (we often find foreign-looking children *more* attractive than our own); indeed for the children of all land vertebrate *species*. We coo over lambs, fawns, newly hatched turtles, even tadpoles!

Our sibling species, chimpanzees, are far more realistic. They too prefer baby animals. They prefer them for food, being more tender. (Humans also have a liking for lamb, calf, piglet, duckling … We can coo over them *and* eat them.) After the kind of warfare now well documented between chimpanzee groups, the victors will kill and eat the young of the vanquished. Male lions will kill the young of prides they take over, and eating the corpses is not unusual. Many mammalian females will eat their young if both are starving, and will frequently 're-process' the first litter in this manner anyway.

No, it's very clear that we are the odd men out. Odd Men indeed. We do have mental circuits to delight in, and protect, our own babies, so that the later Mickey Mouse converged on the outline of a three-year-old toddler, as did E.T. It is no wonder that so many people paid his phone bill. But we also go daft about cuddly juveniles of far too many other kinds. Biologically, this is very odd.

A by-product of our finding other species' babies so attractive has clearly been the domestication of dogs, cats, goats, horses, elephants, hawks, chickens, cattle … These symbioses have given immense pleasure to billions of people, and to their animals, and have contributed greatly to our nutrition. Those who feel that we have exploited the animals unfairly should consider the alternatives for the animals, in the wild, where nearly all of them are eaten alive as babies, without even the benefit of a quick death.

Agriculture can perhaps be attributed to our other propensity, storytelling: the seed becoming the plant has served as an image for so many new words and thoughts, metaphors, new understandings of nature. And the wealth generated by agriculture permitted people to afford princes and philosophers, peasants* and popes. The cultural capital has grown as we have passed our knowledge on to succeeding generations. But it's more comfortable to enjoy that culture if there are a couple of warehouses full of barley for beer, wheat in the fields and cows in the meadows.

We have very recently made the whole business of symbiosis with plants and animals much more technical – those controversial 'genetically modified organisms' – and we have lost a lot by taking our animal helpers out of the system, especially dogs and horses, and replacing them with machinery.

We could not have predicted what the animal and plant symbioses did for us, for our extelligence, and we don't know what losing them will do. Events like that explode on us as the bicycle ride that is extelligence careers down this long technical hill, and can have totally unexpected effects.

Yes, the Ford Model T made motoring affordable to many more people – but a socially much more important change was that it gave privacy in comfort for the first time, so that a large proportion of the next generation was sired on the upholstery of the car's back seat. Similarly, the dog coming in as a symbiont meant that we could hunt more successfully. Then, as a guard dog, it meant that private farms could be protected, and there was help with rounding up the animals and keeping away predators, including human ones. Lapdogs probably affected our sexual courtesies, particularly in eighteenth-century France, and dog and cat shows have stirred our upper-middle classes with, in modern England, the lower aristocracy.

Think for a moment about what we've done to dogs and cats. More than horses and cows, they grow up in our families. We play with them, like we play with our own children, and the play often involves our own children. As with our own children, this interaction begets

* Peasants do cost.

minds in our pets. Even human children don't do much good on the mind front if they don't play. And Jack found, and showed Ian, that even invertebrates, bright invertebrates like mantis shrimps, can acquire minds if they're involved in play activities. We described in *Figments of Reality* how this happens. Let's just note here that we have uplifted* our symbionts into the world of mind. Dogs worry about things, much more than wolves do. So dogs have at least some sense of themselves as a creature that lives in time, with some kind of awareness that it has a future as well as a present. Mind is *catching*.

Usually we think of the domestication of the dog as a selection process driven by human intentions. The process may have started accidentally, perhaps with a tribe bringing up a wolf pup that the kids had brought into the cave, but at a relatively early stage it became a deliberate training programme. Proto-dogs were selected for obedience to their master and for useful skills such as hunting. As time passed, obedience evolved into devotion, and the modern dog arrived on the scene.

However, there is an attractive alternative theory: the dogs selected us. It was the dogs that trained the humans. In this scenario, humans that were willing to allow wolf pups into the cave, and had the ability to train them, were rewarded by the dogs, for example by a willingness to assist in the hunt. Those humans that performed best at these tasks found it easier to acquire new pups and train new generations. The selection on the human side would have been cultural rather than genetic, because there hasn't been enough time for genetic influences to make a lot of difference directly. However, there may well have been selection on the genetic level for having enough intelligence to appreciate how useful a trained wolf could be, or having the generalised teaching skills, such as persistence, to carry out successful training. At any rate, the tribe as a whole benefited from those few individuals who could train proto-dogs, so that the selective pressure in favour of generalised dog-training

* David Brin fans will know what we mean here: in the Five Galaxies, no race (save for the long-defunct Progenitors) ever became extelligent without the aid of a sponsor race which already was. Save for humans, because even in an SF story we need to feel superior. We are, after all, the True Human Beings.

genes must have been slight.

This isn't one of those either-or choices: we are not obliged to accept one theory to the exclusion of the other. And this is a point we should make very firmly, for this and many other theories: things happen, all over the place and apparently in some confusion, and afterwards mankind goes and chops it all up into 'stories'. We need to do it like that that, but occasionally we should stand back and realise what it is we are doing.

In the case of the dogs, there is in all likelihood a fair amount of truth in both theories, and what happened was a complicit co-evolution of dogs and people. As dogs became more obedient and easier to train, people became more willing to train them; as people became more willing to consider possessing a dog, the dogs became more adept at playing along and making themselves useful.

The situation is, perhaps, more clear-cut with cats. Here it was very much a case of the cats being in the driving-seat. Rudyard Kipling's Just-So Story of 'The Cat That Walks By Itself' is too naïve an acceptance of the impression that cats set out to give – that they do things their own way and merely tolerate those people who play along – but in most cases you can't train a cat. Very few cats are willing to perform tricks, whereas many dogs visibly enjoy performing for human pleasure. To the ancient Egyptians, cats were tiny gods on Earth, personified in the cat-goddess Bastet. Bastet was originally worshipped around Bubastis, in the Nile delta, and she had a lion's head; later, this transmuted (transmogrified?) into a cat's head. Bastet-worship then spread to Memphis, where she became conflated with Sakhmet, a local lion-headed goddess. Bastet was a generalised goddess of things that were especially important to women, such as fertility and safe childbirth. Cats were also worshipped, as godling avatars of Bastet, and they were often mummified because of their religious significance. There was a sort of dog-god, the jackal-headed Anubis, but the difference was that he had a substantial 'hands-on' job: he was the god of embalming, and his role was to assist (or impede) the passage of the dead through the underworld. Anubis judged whether the dead were worthy of the afterlife. The only duties that the cat-godlings had were to allow humans to worship them.

Nothing new there, then.

Even today, cats assiduously give the impression of being independent; they seldom come when called, and are liable to depart at a moment's notice for reasons that are never very clear. All cat-owners know, however, that this impression is superficial: their cats need attention, and know it. But this need shows up indirectly. For example, Ian's cat 'Ms Garfield' usually comes out to the front of the house to greet the arrival of the family car, but her pleasure at the car's appearance is heavily disguised as a strident harangue: 'Where the bloody hell have you lot been?' After absences on holiday or overseas, family members find that whenever they are in the garden, the cat coincidentally is in the same part of the garden – but either asleep or apparently just passing through. It looks as though house-cats are slowly losing the domestication battle, but putting up a strong fight. Feral cats are another matter, and real working cats like farm cats are often genuinely independent. These days, though, many farm cats are treated much like house-cats. At any rate, there are some good research projects still to be carried out on the co-evolution of ancient humans and their livestock and pets.

In another instance of this co-evolution, the horse made chivalry a possible culture (hence the name, of course: compare the French 'cheval') and enabled the Mongols to achieve one of the largest, best-controlled empires in human history. Under the Khans it was said that a virgin could walk unmolested from Seville to Hang Chou. Only in the twentieth century was that again achievable, with luck and possibly a harder search for the virgin. The Spanish took horses to America, where humans had killed off several equine species some 13,000 years before,[*] and changed the lives of all the North American Indian tribes – and the cowboys, of course. And, a little later, Hollywood.

The horse did wonders for the genetics of humans, too. Just as they say that the invention of the bicycle saved East Anglia from an

---

[*] Always be careful of the twentieth-century 'story' of 'the natives who live in harmony with their environment'. It tends to gloss over the fact that back in history they killed off all the really big animals, and now it's a choice between harmony and death.

incest implosion, so the people that had come out of Africa were only a tiny part of early *Homo sapiens*'s genetic diversity. All recent studies of the DNA genetics of human populations agree that the genetic diversity outside Africa is only a tiny fraction of the diversity that is still found on that continent. Those who left, to go as far as Australia or China, to Western Europe or via the high Arctic to America, are less diverse in total than many small indigenous African peoples. With the arrival of the horse, it became possible for traders to carry goods – and gene alleles – for very long distances, very effectively. So the out-of-Africa humans have inherited a relatively small part of the African gene-pool: they are genetically impoverished, but well stirred.

At the end of the twentieth century there was, for some years, a belief that *Homo sapiens* was a polyphyletic species. This word means that different groups of *Homo sapiens* evolved from different groups of *Homo erectus* in different places. This, it was thought, might account for the racial differences, especially differences in skin pigmentation, that seemed to fit geography pretty well. From DNA studies, we now know this theory can't be true. On the contrary, there was a bottleneck in our population as we came out of Africa – humanity was reduced to rather small numbers – and all of us living today, all of the out-of-Africa 'races', were extracted from that small population. All the *Homo erectus* died out. The evidence so far looks as if there was only one exodus, of a minimum of some 100,000 people. We were all there *in potentia* in that tiny population, Japanese and Eskimos and Norsemen and Sioux and Beaker people and Mandarin Chinese; Indians and Jews and Irishmen. In the same way, all the current kinds of dog were 'present' in the original domesticated wolf (assuming it was indeed a wolf) – that is, they were in the wolf's space of the adjacent possible – and we've pulled out Saint Bernards and chihuahuas and labradors and King Charles spaniels and poodles from that local region of organism-space.

There was, about thirty years ago, a brief fashion for the concept of 'mitochondrial Eve', and many media reports seem to have picked up the idea that there was just one woman, a veritable Eve, in that ancestral bottleneck. This is nonsense, but the media reports were written up to encourage the belief. The real story, as always, was a little more complicated, and it goes like this. There are mitochondria in the cells of people, indeed of most animals and plants. These are the billions-of-generations, descendants of symbiotic bacteria, and they still have some of their ancient DNA heredity, called mitochondrial DNA. Mitochondria from the mother go into the embryo's cells, but those from father do not: they die, or go only into the placenta. In any event, mitochondrial inheritance is very nearly all maternal. The mitochondrial DNA accumulates mutations over time, with important genes changing less (presumably because the resulting babies, if any, were defective) and some DNA sequences changing quite quickly. That enables us to judge how far back it is to the common ancestor of any pair of women, from the accumulated differences in several DNA sequences. Surprisingly, nearly all such pairs from very different women converge on to a single consensus sequence, about 70,000 years ago.

A single woman, the ancestor of us all.

Eve?

Well, that was the story that the media latched on to, and you can see why. However, it doesn't hang together. The occurrence of just one mitochondrial DNA sequence doesn't mean that there was just one *woman* with that sequence, or that she was the ancestress of all the other women whose DNA was sequenced. Evidence based on the current diversity of various genes shows that there were at least 50,000 women in the human population 70,000 years ago, and *many* of them will have had that particular DNA sequence, or one that cannot be distinguished from it with the evidence remaining today. The lineages of the women who did not have that sequence continued for some time, but eventually died out: their 'branch' of the human family tree doesn't reach all the way to the present day. We can't be certain why those lineages died out, but in mathematical models such effects are commonplace. Perhaps the women carrying sequences like today's sole survivor were more 'fit', or they simply came to outnumber the others by chance. It is even possible that the choice of the contemporary women to test was in some way biased, and that more than one mitochondrial DNA sequence is actually present in today's women.

How do we know that there were at least 100,000 humans 70,000 years ago, and not, as in the stories, just two 6,000 years ago? Many (about 30 per cent) of the genes in the cell nucleus have several versions in today's human population. Like most 'wild' populations (not bred in the laboratory or for dog shows), each individual human has two versions of about 10 per cent of his or her genes, different versions received from father and mother in sperm and egg. Humans have roughly 30,000 genes, of which about 3,000 will be represented by two versions in the average person. For some genes, notably those of the immune system that give each of us a very specific lock-and-key individuality, making us susceptible to some ailments but resistant to others, there are hundreds of versions of *each* gene (of four important ones, anyway). The (common) chimpanzee has a set of these immune variants that is very like the human: in one list of 65 variants of one immune gene, only two were *not* exactly the same. We don't know about the DNA of enough bonobos yet to see if the story is the same for them, but the smart money says that it will be, possibly even more so. The gorilla set seems to be a little different again (but only about thirty gorillas have been tested).

At any rate, *all* of these immune gene variants had to come out of Africa in that 'bottleneck' population that produced all the ex-African human populations. It is unreasonable to suppose that each individual inherited different versions of each variable gene from their parents: some will have carried only one version, the same from both parents, and no one can have carried more than two. The humans that came out of Africa have about 500 immune variants, at least, in common with chimpanzees, out of about 750 possibilities. The humans who stayed in Africa have more: they weren't subject to the bottleneck. There are many other genes where several ancient versions (ancient because they're common to us, chimpanzees, perhaps gorillas, maybe other species) have come through; 100,000 people is a reasonable minimum to carry all those. If you want to be critical and get that number down a bit, you could argue that a few variants from African populations may have been mixed in later, for example via slavery to the US, or to Mediterranean peoples and then via Phoenician sailors to the rest of us. Still, the evidence does not point to an Adam and an Eve, unless they came with a lot of servants, slaves or concubines.

The Biblical stories don't mention those.*

* Mind you, Genesis does say that after Cain killed Abel he was exiled to the land of Nod, on the east of Eden, where he 'knew his wife' and Enoch was born. It doesn't tell us how the wife got to Nod in order to be known. She could have been one of those unmentioned servants, slaves or concubines. That, in turn, raises even more problems with the story of Adam and Eve.

ELEVEN
THE SHELLFISH SCENE

THE WIZARDS WATCHED CAREFULLY.

'There's five of them sitting there with him now,' said Ponder. 'And some children. He seems to be getting on well enough.'

'They're very interested in his hat,' said the Dean.

'A pointy hat always commands respect in any culture,' said Ridcully.

'Then why have several of them tried to eat it?' said the Lecturer in Indefinite Studies.

'At least they don't appear to be warlike,' said Ponder. 'Let's go and introduce ourselves, shall we?'

And, again, when the wizards arrived at the little group around the fire there was the strange sensation of … nothing. No surprise, no shock. The heavy people treated them as if they'd just returned from the bar; their curiosity level extended perhaps to the flavour of crisps they'd brought back, but no further.

'Friendly souls, ain't they?' said Ridcully. 'Which one's the boss?'

Rincewind looked up, and then turned and snatched his hat from a big fist.

'None of them,' he snapped. 'Stop pinching the sequins!'

'Have you mastered their language?'

'I can't! They don't have one! It's all point and kick! That's my *hat*, thank you so very very!'

'We watched you walking around,' said Ponder. 'Surely you've learned something?'

'Oh, yes,' said Rincewind. 'Follow me, and I'll show you – *give me my hat!*'

Holding his sequin-stripped hat firmly on his head with both hands, he led the wizards to a big lagoon on the other side of the village. An arm of the river flowed through it; the water was crystal clear.

'See the shells?' said Rincewind, pointing to a large heap a little way from the beach.

'Freshwater mussels,' said Ridcully. 'Very nutritious. Well?'

'It's a big heap, right?'

'And?' said Ridcully. 'I'm quite fond of mussels myself.'

'You see that hill further along the bank? The one covered in grass? And the one behind that, with all the shrubs and trees? And the – well, see how the whole *area* is a lot higher than rest of the land around here? If you want to know why, just kick the soil away. It's mussel shells all the way down! These people have been here for thousands and thousands of years!'

The tiny clan had followed them and were watching with the uncomprehending interest that was their ground-state expression. Several of them waded in after mussels.

'That's a lot of shellfish,' said the Dean. 'Obviously not a taboo animal.'

'Yes, and that's surprising because frankly these people seem related to them,' said Rincewind wearily. 'Their stone tools are frankly rubbish and they can't build huts and they can't even make fire.'

'But we saw a—'

'Yes. They've *got* fire. They wait for lightning to strike a tree or set fire to grass,' said Rincewind. 'Then they just keep it going for years and years. Believe me, it took a lot of grunting and pointing to work *that* one out. And they have no idea about art. I mean, you know, pictures? I drew a picture of a cow in the dirt and they seemed puzzled. I really think they were just seeing … well, lines. Just lines.'

'Perhaps you're not very good at cow pictures?' said Ridcully.

'Look around,' said Rincewind. 'No beads, no face paint, no decoration. You don't have to be very advanced to knock out a bear claw necklace. Even people who live in caves know how to draw. Ever seen those caves up in Ubergigle? Buffaloes and mammoths as far as the eye can see.'

'I must say you've seemed to strike up a rapport with them very quickly, Rincewind,' said Ponder.

'Well, I've always been good at understanding other people enough to get an inkling of when to start running,' said Rincewind.

'You don't *always* have to run, do you?'

'Yes. Of course. The important thing is to know when it's the appropriate moment, though. Ah, this one's Ug,' said Rincewind, as a white-haired man prodded him with a thick finger. 'So are all the others.'

The current Ug pointed towards the Shell Midden foothills.

'He appears to want us to go with him,' said Ponder.

'He might,' said Rincewind. 'Or he might be pointing out where he last had a really satisfying bowel movement. See them all watching us?'

'Yes.'

'See that strange expression they have?'

'Yes.'

'You wonder what they're thinking?'

'Yes.'

'Nothing. Believe me. That expression means that they're waiting for the next thought to turn up.'

Beyond the Shell Midden Mountains was a thicket of willows, and in the centre of the thicket was a much older tree, or what remained of one. It had been split in two, was now dead, and at some point had been burned.

The clan hung back, but the white-haired Ug followed them into it a little way.

Something crackled under Rincewind's foot. He looked down, saw a yellowing bone, and nearly experienced an appropriate moment. Then he spotted the faint hummocks around the clearing, many of them overgrown.

'And here's the tree that fire came from,' said Ridcully, who had noticed them as well. 'Sacred ground, gentlemen. And they bury their dead.'

'Not exactly *buried*,' said Rincewind. 'More just *left*, I think you'll find. I think they just want to show me where they got fire.'

Ridcully reached for his pipe.

'These people really don't make it?' he said.

'They didn't understand the question,' said Rincewind. 'Well, I *say* question … they didn't understand what I *hope* was the question. We're not talking progressive thinkers here. It must have been a big step when they invented the idea of taking the skins off animals *before* wearing them. I've never met any people quite so … well, *dull*. I can't work them out. They're not *exactly* stupid, but their idea of repartee is an answer within ten minutes.'

'Well, this'll buck their ideas up,' said Ridcully, and lit his pipe. 'I expect they'll be impressed!'

The Ugs looked at one another. They watched the Archchancellor blowing smoke. And then they attacked.

On the Discworld the only tribe known to have absolutely no imaginations whatsoever are the N'tuiftif, although they are gifted with great powers of observation and deduction. They just never invent anything. They were the first tribe ever to *borrow* fire. Being surrounded by other tribes who were as imaginative as anything, they are also very good at hiding; when you are surrounded by tribes to whom a stick means 'club, prod, lever, world domination' you are at a natural disadvantages when, to you, a stick means 'stick'.

To someone else a stick currently meant 'pole'.

A figure vaulted across the clearing and landed in front of the Ugs.

Orangutans do not enter the boxing ring, being too intelligent. If they did, however, the fact that they could knock out the opponent without getting up off their stool would quite make up for lack of finesse in the footwork.

Most of the tribe turned to run, and would have come face to face with the Luggage if it had a face. They rocked when it butted them, and tried to wonder what it was. And by then the Librarian was on top of them.

Those that worked out this was a good time to flee, fled. Those that didn't, stayed on the ground where they had been put.

The astonished Archchancellor was still holding the burning match when the Librarian advanced on him, screaming loudly.

'What say?' he said.

'There's a lot about him being in a library and the next minute being in the river over there,' Ponder supplied.

'That all? Sounded more.'

'The rest was swearing, sir.'

'Apes swear?'

'Yes, sir. All the time.'

There was another burst from the Librarian, accompanied by a pounding of knuckles on the ground.

'More swearing?' said Ridcully.

'Oh *yes*, sir. He's really quite upset. Hex has told him there are no longer any libraries whatsoever at any point in the planet's history.'

'Ow!'

'Quite, sir.'

'I burned my fingers!' Ridcully sucked his thumb. 'Where is Hex, anyway?'

'I was just wondering that, sir. After all, the crystal ball belonged in the city which isn't here any more ...'

They turned and looked at the tree.

It must have blazed furiously when the lightning struck. Probably it had been dead and dry anyway. There were only a few stumps of branches. The whole thing was black, and strangely ominous against the green of the willows.

Rincewind was sitting at the top.

'What the hell are you doing up there, man?' Ridcully bellowed.

'I can't run across water, sir,' Rincewind called down. 'And ... I think I've found Hex. This tree talks ...'

TWELVE
EDGE PEOPLE

 RINCEWIND'S 'EDGE PEOPLE' are a caricature of early hominids, and quite close to what anthropologists used to *think* Neanderthals were like.

We now think that Neanderthals had a bit more going for them, quite apart from burying their dead. At least, it suits the mood of the times to desire to think that they did have something happening behind that big brow-ridge. A bone with holes in it, which some archaeologists believe to be a 43,000-year-old Neanderthal bone flute, has been found in Slovenia. But others dispute that it is a musical instrument. Francesco d'Errico and Philip Chase have studied the bone carefully, and they are certain that the holes were gnawed in it by animals, not bored by a musically minded Neanderthal. We do not know if it's been handed to a musician …

Whatever the status of the flute, it is clear that Neanderthal culture didn't change significantly over long periods of time. The culture that led to us did. It changed dramatically, and so far it's never stopped.

What made us so different from the Neanderthals?

According to the Out of Africa theory, our ancestors, and everybody else's, came from an original population that evolved in Africa. They migrated through the Middle East; the ones bound for Australia probably left from South Africa, but *might* have gone round through the Far East and Malaysia. If you've got boats, you can do either.

In principle the immune-gene story that we discussed in Chapter 10 could tell us more, but nobody's yet done the research: either the

Australian 'aborigines' have the same gene spectrum as the rest of us post-bottleneck humans, or they have their own small and characteristic selection instead. Whichever is the case, it will tell us something interesting, but until someone gathers the genetic data, we have no idea *which* interesting thing it will tell us. A lot of science is like that, a win–win situation. But try explaining that to the bean-counters who control research funding.

When we speak of 'migrations' in this context, you shouldn't think of the exodus of the Hebrews from Egypt. It wasn't a case of one group of humans taking forty years or whatever and conquering other hominids along the way. It was more likely the successive formation of small settlements, slowly getting further and further away from the original homeland. The people themselves didn't even know that they were migrating. It was just 'Hey, Alan, why don't you and Marilyn settle down to hunter-gathering a couple of valleys over, by that nice Euphrates river?' Then, after a hundred years, there would be a few settlements on the far side of the river, too. This isn't pure speculation: archaeologists have found some of the settlements.

If humans formed new settlements a mile away every ten years, it would take only 50,000 years, a mere 1,000 grandfathers, for them to diffuse from Africa all the way to the frozen north. And they surely diffused faster than that. Hardly anybody actually *went* anywhere; it was just that the kids set up home a few hundred yards along the track, where there was a bit more room to bring up *their* kids.

As we diffused, we diversified. It is impressive how diverse we are, physically and culturally. But perhaps, from the elvish viewpoint, we're all much the same, from Chinese to Inuit to Maya to Welshman. Our similarities are far greater than our differences.* We had diversified in Africa, too, from the tall willowy Masai and Zulus to the !Kung†

---

* This is why we have been forced to invent differences of religious belief, which give us an excuse to kill each other because They are so dramatically different from us True Human Beings – they don't even know that spilling salt, and then failing to hop three time around the table, invites a demon into your home. So it's all right to wipe the False Humans, Them, from the face of the planet.

† The ! is a symbol denoting a particular clicking sound.

'pygmies' and the stout Yoruba. These peoples are really, anciently, different: they differ from us, and from each other, almost as much as wolves differ from jackals. The post-bottleneck humans differentiated quite recently, just as the breeds of dogs differentiated from one kind of wolf (or perhaps it was a jackal).

This kind of rapid differentiation is a standard evolutionary story, called 'adaptive radiation'. 'Radiation' means 'spread', and 'adaptive' means that the organisms *change* as they spread, adapting to new environments – and, especially, to the changes brought about by their own adaptive radiation. It happened to 'Darwin's finches', where one small group of finches of a single species arrived in the Galápagos islands, and within a few million years had radiated into 13 separate species, plus a 14th on the Cocos Islands. (We wonder what the legend of *The Fourteenth Finch* might be.) Another well-known example is the vast array of cichlid fish that diversified in Lake Victoria over the last half a million years or so. There they produced variants for the catfish niche, for the planktonic filter-feeder, for the general-detritus feeder; they evolved into species with big crushing teeth for eating mollusc-shells, species that specialised in scraping scales or fins off other fish, and species that specialised in eating the eyes of other fish. Yes, really: when fish from that species were caught, all they had in their stomachs was eyes.* These cichlids ranged in size from a couple of centimetres to half a metre. The original river-dwelling species *Haplochromis burtoni*, whose descendants they all are, grows to a length of 10–12 centimetres.

Curiously, the range of genetics of these fishes was quite small, considering their morphological and behavioural ranges: about the same as out-of-Africa humans, but not as wide as in-Africa humans. At least, that's the case according to some reasonable ways to estimate genetic diversity.

The second part of this story nearly always involves extinction: just occasionally, one of the newly differentiated species has evolved a new and successful trick, and survives while all the others perish.

---

* A meal that should see you through the week, as the old music hall joke reminds us.

But the usual demise of these specialised, adaptively radiated fish happens when a professional specialist – a catfish perhaps, whose ancestors have been feeding on detritus for 20 million years – comes in and takes over from the amateur cichlid catfish. Unfortunately, in this case, it wasn't an inoffensive catfish, but the Nile perch, a specialised carnivore from an ancient stock. The Nile perch has now cleaned out nearly all of the Lake Victoria cichlid explosion, which is why we wrote the previous passage in the past tense.* The main remnants of that glorious radiation of the cichlids are now to be found in the homes of a few amateur hobbyists, who are keeping some of the odd cichlid species in aquaria, and the Geoffrye Museum in London, which by chance has one of the largest ranges of cichlids and is now sponsored by public bodies. We don't know yet if any of the cichlid variants in Lake Victoria has hit on a trick to survive even the Nile Perch.

It's difficult to know what Nile Perch is about to come in and prune *Homo sapiens'* current diversity. With luck, it will be our own propensity to miscegenation, aided and abetted by airlines, despite the contrary admonitions of our priests. Maybe we'll all be mixed up into one fairly diverse type. Or maybe it will be *Independence Day* aliens, out to conquer the galaxy. Or perhaps more competent ones, with elementary virus protection software.

Were we the Nile Perch for the Neanderthals? What was special about us that they couldn't compete with? In an editorial in *Astounding Science Fact and Fiction*, John Campbell Jr proposed that we have been selecting ourselves – in very elvish ways – from earliest times. Campbell credited his idea to the nineteenth-century anthropologist Lewis Morgan, but in truth Campbell contributed most of the story.

It runs: we select ourselves, through puberty rituals and other tribal rites. To some extent these interact with our religious stories, but as a socialising technique the puberty ritual may have preceded all but the most basic of animistic beliefs. It certainly sits at the base of our Make-

---

* Lakes Malawi and Tanganyika still have their cichlid species flocks; your local tropical-fish shop will have representatives.

a-*Homo-sapiens* kit. But the Neanderthals may not have possessed such a cultural kit, at least not in the same effective form. If they didn't, they would probably have been much like Rincewind's edge people, indeed like all the other great apes: settled and (mostly) contented in their Garden of Eden, but not going anywhere.

What is so special about puberty rituals? What story makes them a necessary part of how we evolved ourselves into the storytelling animal? Just this, said Campbell: puberty rituals select the breeders. This is the standard mechanism of 'unnatural selection' used to breed new varieties of dahlias or dogs, only here it bred new varieties of humans or stabilised existing varieties. The wizards have always known about unnatural selection, and it is reified on Discworld as the God of Evolution in *The Last Continent*. Unnatural selection is not just a matter of genetics, either. If you don't get to breed, then you don't have the opportunity to pass on your cultural prejudices to your children. At best you can try to pass them on to other people's children.

Here's how it works. Over there, we see a group of half a dozen lads, perhaps aged 11 to 14. The older men have prepared an ordeal, and the kids must endure this to become accepted as full members of the tribe: that is, breeders. Perhaps they will be circumcised or otherwise wounded, and the wounds will be 'dressed' with painful herbs; perhaps they will be whipped with scorpions or biting insects; perhaps their faces will be seared with red-hot metal brands; perhaps (indeed, usually) the older men will violate them sexually. They will be starved, purged, beaten ... oh yes, we are a very inventive species in this regard.

Those who ran away were not accepted into the group,* and so were not breeders. So, in particular, they were not our ancestors, because they weren't anyone's ancestors. In contrast, those who submitted to the humiliation were rewarded by acceptance into the tribe. Campbell's insight was that these puberty rituals selected *against* the immediate animal avoidance-of-pain response, and selected *for* both imagination and heroism: 'If I bear this pain now I will be rewarded

---

* 'Going walkabout' seems to have been a way to avoid this torture for at least some Australian tribes.

by getting the privileges these old men get, and I can imagine that they went through exactly this, and survived.'

Later on it was the priests who administered the pain. That is how they *became* the priests, and how successive generations came to 'respect' them and their teachings. By then, humiliation had become its own reward, at both ends of the instrument of torture (see *Small Gods*), and humans had been selected for obedience to authority.

Indeed, Stanley Milgram's book *Obedience to Authority* shows just how obedient we are, by using the authority of a white laboratory coat to force people to torture other people, remotely. The other people were actually actors, responding to 'mild', 'strong' and 'excruciating' pain – or so the experimental subject was led to believe – with the appropriate actions. Milgram's book shows how human beings invented authority and obedience, both very elvish sentiments. That ingredient in the story of our evolution explains Adolf Eichmann as well as Einstein: we won't go any deeper into that issue here, because we've already covered it in *The Privileged Ape* and *Figments of Reality*.

A few people refused Milgram's instructions, though, and these mavericks have always been generated either by experience (some of the refusers had survived concentration camps, or had been otherwise tortured themselves) or by the Make-a-Human kit itself. Many of these kits generate a few mavericks, and we are optimistic about the Western one that uses Hollywood films to laud resistance to authority. But perhaps that comes only by working through the right genetics and the right home background.

Many of these ancient rituals have become empty now. Jews use circumcision to test the parents' commitment, rather than that of the baby, who has no choice. Jack was the Boston foreskin collector in the early 1960s; it was a very good source of the living human skin samples that he needed for his research on pigment cells in the skin. He saw a lot of parents, many of whom went very pale and a few of whom fainted: more men did that than women. The Jewish Bar Mitzvah is very daunting to the child, in prospect, though, as with circumcision, nobody *fails* it – not any more. But people did fail in the past, with serious consequences. For example in the ghettos, where only a third of the population married, the mothers of the 'best' girls chose only the boys who performed their Bar Mitzvah best. This would account for the kind of verbal success that the Jewish faction of many Western populations has achieved. Another explanation, that Jews were permitted verbal abilities only because land- or property-owning was denied them, is a contextual constraint within which they had to live. Why they were good enough verbally to succeed despite that constraint is the interesting question, and Bar Mitzvah competition and selection of breeders is a persuasive answer.

Gypsy populations provide a possible counterexample, though, with very little testing of young men before marriage, which frequently takes place at ages that other cultures consider to be pre-pubertal. The few gypsies who have been successful in Western cultures have not been primarily verbally successful. Music provides a good contrast, with gypsies excelling in dance while classical composers and instrumental soloists are often Jewish. Of course, gypsies also share our common selective ancestry, if we're right about puberty rites being ancestral and effectively universal.

The other great apes don't torture their children for ritual purposes, and the other hominids like Neanderthals probably didn't either. So they haven't produced a civilisation. Sorry, but that which does not kill us *does* appear to have made us strong.

There is another story that we now tell, about what happened to the young men around the time when people were inventing agriculture, which explains barbaric societies. Don't get us wrong here: we don't mean that torturing adolescents is barbaric. It's not, from the tribal point of view. It is an entirely proper way to get them accepted into the tribe. 'We've done it ever since god-on-high made the world, and to prove it, here's the holy circumcision-knife we've always used.' No, from the tribal point of view, the barbarians that we have in mind are awful; they don't have *any* rules or traditions … Even the Manky tribe, over that way a couple of miles, is better than them; at least the Mankies *have* traditions, even if they are different from ours. And we've stolen some of their women, and they have the most *amazing* tricks …

The problem is that lot up on the hillside, the young men who have been expelled from the tribe because they failed the rituals, or went of their own accord (and so failed the test anyway). 'Couple of my brothers up there with 'em, and Joel's boy, and of course the four kids that were left when Gertie died. Oh, they're all right on their own; it's when they're in that gang together, all doing their hair in that same funny way to be different, that you lock up the sheep and let the dogs loose. They've got these funny words like "honour" and "bravery" and "pillage" and "hero" and "our gang". When my brothers come down the valley to my farm – by themselves – I give 'em some food. But some gang of young men, I'm not saying it was that lot and I'm not saying it wasn't, set the Brown's farm alight, just for the hell of it …'

In any cowboy film we find the message that barbarism is opposed to tribalism, that honour and tradition are not good bedfellows. And that, having selected himself or herself for imagination and the ability to endure pain for future pleasure, *Homo sapiens* is now prepared to die for his or her beliefs, for his or her gang, for honour, for hatred, or for love.

Civilisation, as we know it, seems to combine elements of both ways of human culture, tribal by tradition and barbaric for honour, for pride. Nations are internally tribal, but present a barbarian face to other nations. Our extelligence tells us stories, and we tell our children stories, and the stories guide us about what to be or do in what circumstances. Shakespeare is the ultimate civiliser, in this view. His plays were composed against the barbarian background, in a city where you could see heads on spikes and ritually dismembered bodies; all of them were set on the tribal, traditional base that is most of human life, most of the time. He tells us very persuasively that evil fails in the end, that love conquers, and that laughter – the greatest gift that barbarism brought to tribalism – is one of the sharpest weapons, because it civilises.

Cohens are the hereditary High Priest lineage of the Jews. Jack was once asked, in Jerusalem, whether he was not proud to be a Cohen, in view of the noble Jewish history that the High Priests had promoted. Jack sees this nobility as based in about six inches of blood in the streets, nearly all of it other people's, so he is not proud. Instead, to the extent that any of us is responsible for what their ancestors did, he is ashamed. He loves *Small Gods*, in much the way he enjoys the Jewish Day of Atonement, Yom Kippur: it engenders a feeling of repentance, and he can always find plenty to repent. He is sure that this emotion – guilt – is a legacy of the Morgan/Campbell selection of his ancestors through tribal rituals.

Tribesmen aren't 'proud'; for them, everything that isn't mandatory is forbidden, so what is there to be proud about? You can praise your children for doing things right, or admonish or punish them for doing things wrong, but you can't take pride in what you – a fully fledged member of the tribe – do. That comes with the territory. However, you *can* be guilty about not having done the things that you should have done. Having said that, High Priests waging war on dissenters or neigh-bouring tribes, leading to atrocities like heads on spikes, is straight barbarism.

The distinction between tribalism and barbarism is illuminated by the story of Dinah in chapter 34 of Genesis. Dinah, an Israelite, was the daughter of Leah and Jacob, and 'when Schechem the son of Hamor the Hivite, prince of the country, saw her, he took her, and lay with her, and defiled her'. Then Schechem fell in love with her, and wanted to make her his wife. But the sons of Jacob felt that maybe Schechem had gone about things in the wrong order: '... the men were grieved, and they were very wroth, because he had wrought folly in Israel in lying with Jacob's daughter, which thing ought not to be done'. So when Hamor, the father of Schechem, asked for approval of the marriage, and for an intermingling of his tribe with the Israelites, the sons of Jacob came up with a cunning plan.

They told the Hivites that they would agree to the proposal, but only after the Hivites had circumcised themselves, so that they were just like the Israelites. The Hivites were willing to go along with this, because they told themselves that 'These men are peaceable with us, therefore let them dwell in the land, and trade therein; for the land, behold, it is large enough for them; let us take their daughters to us for wives, and let us give them our daughters'. The decision was made, and 'every male was circumcised, all that went out of the gate of the city'. And they stood around in pain for a couple of days. At that point,

Dinah's brothers Simeon and Levi hauled Dinah out of Schechem's house, put all the Hivite men to the sword, destroyed their city and took all their domestic animals, their wealth, their children and their wives. This story of deceit and betrayal has not been given much circulation in recent years; it doesn't appeal to people's sense of humour any more, as it once did.

At any rate, in that story, the Hivite response to Schechem's crime is tribal, but the Israelites behave like barbarians. The Hivites, after their initial mistake, want to make amends and coexist peacefully, and they're prepared to offer dowries and other concessions to try to make up for what Schechem did. But all that matters to the Israelites is a twisted kind of 'honour', in which cruelty, murder and theft are justified to protect Dinah's reputation. Or, more likely, their own sense of manhood.

A favourite Discworld character is Cohen the Barbarian, a satire on sword-and-sorcery heroes like Conan the Barbarian, all muscles and trolls' teeth necklaces and testosterone-propelled heroism. He first appears in the second Discworld novel *The Light Fantastic*:

> 'Hang on, hang on,' said Rincewind. 'Cohen's a great chap, neck like a bull, got chest muscles like a sack of footballs. I mean, he's the Disc's greatest warrior, a legend in his own lifetime. I remember my grandad telling me he saw him ... my grandad telling me he ... my grandad ...'
>
> He faltered under the gimlet gaze.
>
> 'Oh,' he said. 'Oh. Of course. Sorry.'
>
> 'Yesh,' said Cohen, and sighed. 'Thatsh right, boy. I'm a lifetime in my own legend.'

Cohen, by then 87, is the sort of barbarian whose hordes ride into town, set the houses on fire and look wistfully at the women. But he's no softie: as he ages, he goes hard, like oak. In *Interesting Times* he explains to Rincewind why, in the area known as the Ramtops, there's no future in the Barbarian business any more:

'Fences and farms, fences and farms *everywhere*. You kill a dragon these days, people *complain*. You know what? You know what happened?'

'No. What happened?'

'Man came up to me, said my teeth were offensive to trolls. What about that, eh?'

According to Jewish tradition, Cohens are the true Cohanim, the lineal descendants of Aaron. Recent research into the genetics of Cohens has turned up some interesting findings about the very prideful (barbaric) issue of Cohen heredity. Professor Vivian Moses (yes, indeed …) and a group of scientists in Israel decided to check whether the tradition has any factual basis. Just as the mitochondrial DNA sequence traces female heredity, so the Y-chromosome, possessed only by males, can be used to trace male heredity.

There has been an interesting division of the Jewish peoples, and that provides a scientific check on the story of the Cohanim. During the Diaspora, some Jews remained in North Africa, but one large population went into Spain. They are known as Sephardi, and the Rothschilds, Montefiores and other banking families are all Sephardic. Another, more diffuse population went into middle-Europe, especially Poland, and they are known as Ashkenazi. Moses and his colleagues looked at the Y-chromosomes of representative Sephardi and Ashkenazi Cohens and non-Cohens ('Israelites'). They found characteristic DNA sequences, specific to Cohanim, in about half of the Cohens that they tested, but with small and characteristic differences in the three groups. From these differences it is reasonable to suppose that Ashkenazi and Sephardi Jews separated rather less than 2,000 years ago, and that all Cohens were a single group only 2,500 years ago.

This looks like a very nice story, with the DNA evidence supporting the expected history. But science is the best guard against believing things because you want to. There is a factor that Moses and his colleagues didn't explicitly consider, and it needs to be explained away, because it makes those figures much too good.

Most human groups pretend to practise monogamy, but like swans and gibbons and other creatures that we thought were faithful for life, there are plenty of adulterous relationships and children 'whose legal and biological parentage differ'. In English society, about one child in seven is in that position, and the proportion doesn't differ much between the slums of Liverpool and the stockbroker belt of Maidenhead.*

The most restrained people that we know, in this regard, are the Amish of Eastern Pennsylvania and other parts of the United States, for whom the figure is a mere one in twenty. So, to err on the safe side, let's assume that all of the Mrs Cohens, from the present day back 100 generations to the sons of Aaron, were as well behaved as the Amish. Then the proportion of Cohen males with Aaron's Y-chromosome should be 0.95^{100}, which is considerably less than one in a hundred. So how can it be as high as one in two?

There is a possible explanation, consonant with what we know about human sexuality, or at least with what John Symons, an expert on human sexual practices, says in his books. According to many surveys of sexual behaviour, going right back to Alfred Kinsey around 1950, women practise adultery with men of both higher and lower status. The two situations frequently occur in different social contexts, with women 'doing favours for' higher status men (think Clinton), but going down-market for 'a bit of rough'. Overwhelmingly often, however, when a baby results the father is of higher status than the woman's husband or regular partner.

This implies that if Mrs Cohen, living in a ghetto or any other pre-

* Yes, we know you don't believe this, but ... The first reliable data are in Elliott Philipp's analysis of blood-groups from families in high-rise apartments in Liverpool in the late 1960s, published in 1973. There, 10 per cent of the 'legal paternities' were biologically impossible. So, correcting for the cases where the milkman had the same blood-group as the legal father, about 13–17 per cent were 'discrepant paternities', as the coy phrase goes. Hundreds of births in Maidenhead, in the stockbroker-belt, yielded the same proportions. American figures for the 1980s were about 10 per cent, but these were underestimates because they were not corrected as above. That's the thing about science: it tells you stuff you didn't expect. It gets worse. Or maybe you feel it gets better. At any rate many animals that until recently were famed for their fidelity, such as swans, turn out to be partial to a bit on the side. That ubiquitous beast, the monogamus, is rapidly going extinct.

dominantly Jewish society, wants to go up-market, her only choices are other Cohens. So the maintenance of the Aaron Y-chromosome may have been assured by sexual snobbery rather than amazing fidelity, and that's a much more likely story.

THIRTEEN
STASIS QUO

 THE BREEZE SHOOK THE WILLOWS. And, in the centre of the willows, the lightning-struck tree spoke, in a very faint voice. The Ugs had seen lightning strike the same tree three times. It was the highest point in the area, thanks to the shell mounds.

Even for creatures so preternaturally against thinking any kind of new thoughts, this made an impression. In some way, they'd felt, the tree had importance. It was an important thing. The place of the tree was an important place, where the sky touched the ground.

It wasn't much of an opening, it was a story without a plot and it barely amounted to a belief, but Hex had to make do with what could be found.

Now the wizards were considering the future, or futures.

'*Nothing* changes?' said the Dean.

'No, sir,' said Ponder, for the fourth time. 'And, yes, this is indeed the same *time* as the city we were in. But things are different.'

'The city was almost modern!'

'Yes, it had heads on spikes,' said Rincewind.

'It was a bit backward, admittedly,' said Ridcully. 'And the beer was foul. But it had possibilities.'

'But I don't understand! We *stopped* the elves,' said the Dean.

'And now we've got thousands and thousands of years of this,' said Ponder. 'That's what Hex says. These people won't even have learned how to make fire before the big rock hits. Rincewind is right. They're not exactly stupid, they just don't ... progress. Remember the crab

144

civilisation we found?'

'But they had wars and took prisoners and slaves!' said the Lecturer in Recent Runes.

'Yes. Progress,' said Ponder.

'Heads on spikes,' said Rincewind.

'Do stop going on about that, it was only two heads,' snapped Ponder.

'Perhaps we did something else that changed history,' said the Chair of Indefinite Studies. 'Maybe we trod on the wrong insect or something? Only a thought,' he added, when they glared at him.

'We just saw off the elves, that's all we did,' said Ridcully. 'Elves cause exactly the sort of things we've seen here. Superstition and—'

'The Ugs aren't superstitious,' said Rincewind.

'They didn't like it when I struck that match!'

'They didn't start worshipping you, either. They just don't like things that happen too quickly. But I *told* you, they don't draw pictures, they don't use body paint, they don't *make* things … I asked Ug about the sky and the moon, and as far as I can tell they don't think about them. They're just things in the sky.'

'Oh, come now,' said Ridcully, '*everyone* tells stories about the moon.'

'They don't. They don't have any stories at all,' said Rincewind.

There was silence as this sank in.

'Oh dear,' said Ponder.

'No narrativium,' said the Dean. 'Remember? That's what this universe lacks. We never found a trace of it. Nothing knows what it's supposed to be.'

'There must be *something* like it, surely?' said Ridcully. 'The place looks normal, after all. Seeds grow up into trees and grass, by the look of it. Clouds know they have to stay up in the sky.'

'If you remember, sir,' said Ponder, using the tone that meant *I know you've forgotten, sir*, 'we found that this universe has things that work instead of narrativium.'

'Then why are these people just sittin' about?'

'Because that's all they have to do!' said Rincewind. 'There doesn't seem to be much around that can hurt them, there's enough food, the sun is shining … it's all gravy! They're like … lions. Lions don't need stories. Eat when you're hungry, sleep when you're tired. That's all they need to know. What else do they need?'

'But it must get cold in the winter, surely?'

'So? It gets warmer in the spring! It's just like the moon and the stars! Things happen!'

'And they've been like this for hundreds of thousands of years,' said Ponder.

There was some more silence.

'Remember those stupid big lizards?' said the Dean. 'They lasted for more than a hundred million years, I remember. I suppose they were quite successful, in their way.'

'Successful?' said Ridcully.

'I mean they lasted a long time.'

'Really? And did they build a single university?'

'Well, no—'

'Did they draw a single picture? Invent writing? Offer even small classes of elementary tuition?'

'Not that I know—'

'And they all got killed off by a yet another big rock,' said Ridcully. 'They really did not know what hit them. Bein' around for millions of years is not an achievement. Even lumps of stone can manage that.'

The circle of wizards was sunk in gloom.

'And Dee's people were doin' quite well,' muttered Ridcully. 'Terrible beer, of course.'

'I suppose …' Rincewind began.

'Yes?' said the Archchancellor.

'Well … how about if we went back and stopped us from stopping the elves? And least we'd be back among people more interesting than cows.'

'Could we do that?' said Ridcully to Ponder.

'I suppose so,' said Ponder. 'Technically, if we stop ourselves, then nothing will change, I assume. All this won't have happened … I think. That is to say, it *will* have happened, because we'll remember it, but then it won't have happened.'

'Fair enough,' said Ridcully. Wizards do not have a lot of patience with temporal paradoxes.

'*Can* we stop ourselves?' said the Dean. 'I mean, how do we do it?'

'We'll just explain the situation to us,' said Ridcully. 'We're reasonable men.'

'Hah!' said Ponder, and then looked up. 'Oh, sorry, Archchancellor. I must have been thinking about something else. Do go on.'

'Ahem. If I was just about to fight elves, and someone who looked very much like me came up and told me not to, I'd assume it was an elvish trick,' said the Lecturer in Recent Runes. 'They can make you think they look like someone else, you know.'

'I'd know me if I saw me!' said the Dean.

'Look, it's easy,' said Rincewind. 'Trust me. Just tell yourself something about yourself that no one else could possibly know.'

A worried look crossed the Dean's face.

'Would that be wise?' he said. Like many people, wizards often have secrets they don't want themselves to know.

Ridcully stood up. 'We know it'll work,' he said, 'because it's already happened to us. Think about it. We must succeed in the end, because we know a species like this gets off the planet.'

'Yes,' said Ponder, slowly, 'and, then again, no.'

'What the hell does that mean?' Ridcully demanded.

'Well … we've been to a future where it happens, certainly,' said Ponder, twiddling his pencil nervously. 'But there are other futures. The multiplex nature of the universe that allows it to absorb and cushion the effects of apparent paradoxes also means that nothing is certain, even if you know it is.' He tried to avoid Ridcully's stare. 'We went to a future. At the moment, it exists only in our memories. Then, it was real. Now, it may never be. Look, Rincewind was telling me about some play writer he's found, born around about Dee's time but not in this branch of the universe. Yet we know he has *an* existence, because L-space contains all possible books in all possible histories. Do you see what I mean? Nothing is certain.'

After a while, the Chair of Indefinite Studies said, 'You know, I think I prefer the kind of universal law that says the third son of a king always gets the princess. *They* make sense.'

'The universe is so big, sir, that it obeys all possible laws,' said

Ponder. 'For a given value of "teapot".'

'Look, if we go back in time and talk to ourselves, why don't we remember it?' said the Lecturer in Recent Runes.

Ponder sighed. 'Because although it has already happened to us, it hasn't yet happened to *us*.'

'I, er, tried something like that,' said Rincewind. 'While you were having your mussel soup just now I got Hex to send me back in time to warn myself to hold my breath when we landed in the river. It worked.'

'Did you hold your breath?'

'Yes, because I've warned myself.'

'So … was there any time anywhere where you didn't hold your breath, thus giving yourself a mouthful of river water and causing you to go back to make sure you *did*?'

'Probably there was, I think, but there isn't now.'

'Oh, I *see*,' said the Lecturer in Recent Runes. 'You know, it's a good job we're wizards, otherwise this time travel business could really be confusing …'

'At least we know that Hex can still make contact with us,' said Ponder. 'I'll ask him to move us back again.'

The Librarian watched them go.

A moment later, the rest of everywhere went with them.

FOURTEEN
POOH AND
THE PROPHETS

THE UGS HAVE NO REAL STORIES, hence no sense of their place in time. They have no conception of the future, and therefore no wish to change it.

We know that there are other futures …

As Ponder Stibbons remarks, we live in a multiplex universe. We look at the past and we see times and places where things could have been different, and we wonder whether we could have ended up in a different present. By analogy, we look at the present and imagine many different futures. And we wonder which of them will happen, and what we can do now to affect the choice.

We could be wrong. Maybe the fatalist view, 'it is written', is right. Maybe we are all automata, working out the deterministic future of a clockwork universe. Or maybe the Quantum philosophers are right, and all possible futures (and pasts) coexist. Or maybe everything that exists is just one point in a multiplex phase space of universes, a single card dealt from Fate's deck.

How did we acquire this sense of ourselves as beings who exist in time? Who remember their past, and use it to try (usually unsuccessfully) to control their future?

It all goes back a long, long way.

Watch a proto-human watching a zebra watching a lioness. The three mammalian brains are doing very different things. The herbivore brain has seen the lioness, is barely conscious (we guess, watch some horses in a field) of the whole 360 degrees of his environment, and has marked some things, like that tuft of grass over there, that female over there who could just be in heat, that male who's giving her the right signals, the three bushes that could have a surprise behind them ... If the lioness moves, she suddenly gets priority, but not totally because there are other considerations. Another lioness could well be behind those bushes, and I'd better move up on that nice grass before Nigella does ... Looking at that grass makes me think of the taste of that long grass ... THE LIONESS IS MOVING.

The lioness is thinking: that's a nice zebra stallion, won't go for him, he's too strong (memory of a previous eye injury from a zebra kick), but if I get him running, Dora behind those bushes can probably jump on the young female over there who is trying to attract the male, then I can run after it with her ...

There is probably no more of a plan than that in the zebra's brain, but it does foresee a little bit of the future and plug memories into present planning. *If I stand up now* ...

The human is looking at the lioness and the zebra. Even if it's a *Homo erectus*, we bet it had stories in its head: that lioness will run out, the zebra will startle, the other lioness will go for ... ah, that young female. Then I can run out there and get in front of the young male; I see myself running at him and hitting him with this stone. *Homo sapiens* may well have done better from the beginning; his brain was bigger and probably better. He may, from the beginning, have had room for several alternative, thought-about 'or' scenarios and probably the 'and' one which goes 'and I will be a big hunter and meet interesting women'. 'If' probably came along later, perhaps with cave paintings, but making predictions put our ancestors way ahead of their predators and their prey.

There has been a variety of suggestions about why our brains suddenly grew to nearly double their previous size, from the need to keep the faces of our social group in mind while gossiping about them, to the need to compete with other hunter-gatherers, to the competitive nature of language and its structuring of the brain so that lying could be successful for the li-ar, but then the li-ee got better at detecting lies. Such escalations all have an attraction to them. They make good stories, ones that we can easily imagine, filling in the background just as we do with hearing sentences or enjoying pictures. That doesn't make them true, of course, just as our attraction to the supposed seashore phase of our history doesn't make 'aquatic apes' true either. The stories serve as placeholders for whatever the real pressures were: the meta-explanation of why our brain growth took off is that competitive advantage was to be won by All Of The Above routes, and many more.

Perhaps the human viewer of that wildlife scene is a cameraman for a natural history TV series. Even a mere 15 years ago, he would have had an Arriflex (or if he was paying for it himself, perhaps just a Bolex H16) 16mm film camera with a very precious 800 feet (260 metres) of film loaded, and perhaps another dozen film packs in his rucksack (800 feet gives about 40 minutes of filming: if you're very good, or very lucky, five minutes of useful stuff). *Now* he has a video camera that would have seemed miraculous then, which can reuse and reuse a length of tape until it's full of five-minute sequences, end to end. All the things he wished for, then, are in the apparatus in his hand now: it stays in focus, it compensates for a bit of wobble, it goes down to *unbelievably* low light levels (for those of us who grew up with photographic film) and it zooms over a range much wider than we ever had before.

It's magic, in fact.

And in his head are a dozen alternative scenarios for the lions and zebras, which he'll flick to instantly as the animals act to constrain their futures. He's actually thinking about other things altogether, letting the experienced professional part of his brain do the work while he daydreams ('I'll get an award for this and meet interesting women'). It's like driving on a quiet motorway: a lot of the thinking has been taken out of it.

Our ancestors honed that ability, to consider alternative scenarios. And within any of those scenarios, the ability to make a story of what was happening was a very powerful way to remember it and to communicate it. And, particularly, to employ it as a parable, to direct your future action or that of your children. Human beings need a very long time to get that brain up-and-running, at least twice as long as their brother and sister chimpanzees. That is why three-year-old chimps are nearly adult in chimp behaviour, and can do some of the mental tricks of six- or seven-year-old children.

But the young chimps don't hear stories. Our children have been hearing stories since they recognised any words at all, and by three years old they are making up their own stories about what is happening around them. We are all impressed by their vocabulary skills, and by their acquisition of syntax and semantics; but we should also note how good they are at making narratives out of events. From about five years old, they get their parents to do things for them by placing those things in narrative context. And most of their games with peers have a context, within which stories are played out. The context they create is just like that of the animal and fairy stories we tell them. The parents don't instruct the child how to do this, nor do the children have to elicit the 'right' storytelling behaviours from their parents. This is an evolutionary complicity. It seems very natural – after all, we are *Pan narrans* – that we tell stories to children, and that children and parents enjoy the activity. We learn about 'narrativium' very early in our development, and we use it and promote it for the whole of our lives.

Human development is a complex, recursive behaviour. It is not simply reading out DNA 'blueprints' and making another working part (contrary to the new folk-biology of genes). To show you how truly remarkable our development is, despite seeming so simple and so natural, let's have a look at some earlier parent–child behaviour.

Keep in mind a distinction that is being imported into more and more scientific thinking, that between 'complicated' and 'complex'. 'Complicated' means a whole set of simple things working together to produce some effect, like a clock or an automobile: each of the components – brakes, engine, body-shell, steering – contributes to what the car does by doing its own thing, pretty well. There are some interactions, to be sure. When the engine is turning fast, it has a gyroscopic effect that makes the steering behave differently, and the gearbox affects how fast the engine is going at a particular car speed. To see human development as a kind of car assembly process, with the successive genetic blueprints 'defining' each new bit as we add them, is to see us as only complicated.

A car being driven, however, is a *complex* system: each action it takes helps determine future actions and is dependent upon previous actions. It changes the rules for itself as it goes. So does a garden. As plants grow, they take nutrients from the soil, and this affects what else can grow there later. But they also rot down, adding nutrients, providing habitat for insects, grubs, hedgehogs … A mature garden has a very different dynamic from that of a new plot on a housing estate.

Similarly, we change our own rules as we develop.

There are always several superficially different, non-overlapping descriptions of any complex system, and one way to deal with a complex system is to collect these descriptions and choose appropriate ones for different ways of influencing its behaviour.* An amusingly simple example can be seen in many French and Swiss railway stations and airports: a sign that says

LOST PROPERTY

OBJETS TROUVÉS

The French means '*found* objects'. But we don't think that this is a case of the English losing objects and the French finding them. It's two descriptions of the same situation.

Now look at a baby in a pram, throwing its rattle out on to the pavement for Mummy, or child-minder, or indeed passers-by, to retrieve. We probably think that the child is not coordinated enough yet to keep its rattle within reach: we think 'Lost Property'. Then we see Mummy give the rattle back to the child, to be rewarded with a smile, and we think 'No, it's more subtle: there is a baby teaching its mother to fetch, just as we adults do with dogs'. *Now* we think 'Objets Trouvés'. The baby's smile is itself part of a complex, reciprocal system of rewards that was set up long ago in evolution. We watch babies 'copy' the smiles of parents – but no, it can't be copying, because even

---

* Until we had really good fast computers, and had learned a little bit about how to model the complexity of ecosystems or companies or bacterial communities, most of us practised the reductionist trick of looking for the bits we thought we could understand and modelling those. Then we hoped we could put these separate bits together to understand the whole thing. We were nearly always wrong.

blind babies smile. Anyway, copying would be immensely difficult: from anywhere on the retina, the undeveloped brain must 'sort out' a face with a smile, then work out which of its own muscles to work to produce that effect, without a mirror. No, it's a pre-wired reflex. Babies reflexly react to cooing sounds and to pre-wired recognition of smiles; an upwardly-curved line on a piece of paper works just as well. The 'smile' icon rewards the adult, who then tries hard to keep the baby doing it. The complex interactions proceed, changing both participants progressively.

They can be analysed more easily in unusual situations, such as sighted children with 'signing' parents, perhaps deaf or dumb, but occasionally as part of a psychological experiment. For example in 2001 a team of Canadian researchers headed by Laura Ann Petitto studied three children, about six months old, all with perfect hearing but born to deaf parents. The parents 'cooed over' the babies in sign language, and the babies began to 'babble' sign language – that is, make a variety of random gestures with their hands – in return. The parents used an unusual and very rhythmic form of sign language, quite unlike anything they would use to adults. Similarly, adults speak to babies in a rhythmic sing-song voice, and between the ages of about six months and a year the babies' babble takes on properties of the parents' specific language. They are rewiring and 'tuning' their sense organs, in this case the cochlea, to hear that language best.

Some scientists think that babbling sounds is just random opening and closing of the jaw, but others are convinced it is an essential stage in the learning of language. The use of special rhythms by parents, and the spontaneous 'babbling' with hand-movements when the parents are deaf, indicate that the second theory is closer to the mark. Petitto suggests that the use of rhythm is an ancient evolutionary trick, exploiting the natural sensitivities of the young child.

As the child grows, its complex interaction with surrounding humans comes to produce wholly unexpected results: what we call 'emergent' behaviour, meaning that it is not overtly present in the behaviour of the components. Where two or more systems interact like this, we call the process a *complicity*. The interaction of an actor with an audience can build up a wholly new and unexpected relationship. The evolutionary interaction of blood-sucking insects with vertebrates paved the way for protozoan blood parasites that cause diseases like malaria and sleeping-sickness. The car-and-driver behaves differently from either alone (and car-and-driver-and-alcohol is even less predictable). Similarly, human development is a progressive interaction between the child's intelligence and the culture's extelligence: a complicity. This complicity progresses from simple vocabulary-learning to the syntax of little sentences and the semantics of fulfilling the child's needs and wants and the parents' expectations. The beginning of storytelling then becomes an early threshold into worlds that our kin the chimpanzees know not of.

The stories that all human cultures use to mould the expectations and behaviour of the growing child use iconic figures: always some animals, and then status-figures of the culture (princesses, wizards, giants, mermaids). These stories sit in all our minds, contributing to our acting, our acting-out, our thinking, our predicting what will happen next, as caveman or cameraman. We learn to expect outcomes of particular kinds, frequently expressed in ritual words ('And they all lived happily ever after' or 'So it all ended in tears').* The stories that have been used in England over the centuries have changed in complicity with the changing culture – making the culture change, and responding to those changes, like a river changing its path across a wide flood-plain that it has itself built. The Grimm Brothers and Hans Christian Andersen were but the last of a long series, with Charles Perrault accumulating the Mother Goose tales around 1690; there were many collections before that, especially some interesting Italian groupings and retellings for adults.

The great advantage we all get from this programming is very clear. It trains us to do 'What if …?' experiments in our minds, using the rules that we've picked up from the stories, just as we picked up syntax by

---

* As G.K. Chesterton pointed out, fairy tales are certainly not, as modern detractors of the fantasy genre believe, set in a world 'where anything can happen'. They existed in a world with rules ('don't stray from the path', 'don't open the blue door', 'you must be home before midnight', and so on). In a world where *anything* could happen, you couldn't have stories at all.

hearing our parents talking. These stories-of-the-future enable us to set ourselves in an extended imagined present, just as our vision is an extended picture reaching much further out in all directions than the tiny central part to which we're paying attention. These abilities enable each of us to see ourselves as being set in a nexus of space and time; our 'here' and 'now' form only the starting place for our seeing ourselves in other places at other times. This ability has been called 'time-binding', and has been seen as miraculous, but it seems to us that it is the culmination (for now) of an entirely natural progression that starts from interpreting and enlarging vision or hearing, and from 'making sense' in general. The extelligence uses this faculty, and hones and improves it for each of us, so that we can use metaphor to navigate our thoughts. Pooh Bear getting stuck, and unable to exit with dignity because he ate too much honey, is precisely the kind of parable that we carry with us to guide our actions, as metaphor, from day to day. So are Biblical stories, with all their lessons for life.

Holy books like the *Bible* and the *Koran* take this ability one giant step further. The Biblical prophets do, wholesale, what each of us has been programmed to do retail for our own life and those of our own nearest and dearest. The prophets predicted what would happen to everybody in the tribe if they continued their current behaviour, and thereby changed that behaviour. This was a step on the way to those modern prophets who predict The End Of The World some time soon. They seem to feel that they have perceived a trend, a constraint in the universe, that the rest of us have not understood, and whose properties are directing the universe along some undesirable or calamitous path. Though they don't usually mean 'universe', they mean 'my world and nearer ones'. So far they haven't been right. But we would not be here to write these words if they had been, which is another anthropic issue, but not a very important one because they *have* been wrong rather often. They predict what will happen If This Goes On; but, increasingly it seems, This doesn't Go On for very long because it's unexpectedly replaced by a new This.

We all think that we can become better prophets with practice. We all think we have a clever way to build 'the road not taken' into our experience. Then we invent time travel, at least in our imaginations. We all want to go back to the beginning of that argument with the boss, and do it right this time. We want to unravel the chain of causality that led to boring edge people. We want to avoid the bad effects of elves but retain the good ones. We want to play pick-and-mix with universes.

However, despite their emphasis on prophecy, monotheist faiths have real trouble with multiple futures. Having simplified their theology down to one God, they also tend to believe that there can be only one 'right way to heaven'. The priests tell the people what they must do, and at least while the religion is fresh the priests are fine examples. This is what gets you to heaven, they say: no adultery, no murder, no failure to give a tithe to the Church, and no undercutting the other clergy for indulgences. Then the gateway to heaven becomes 'strait', narrower and narrower, until only the blessed and the saints can get in without spending time in some purgatory or other.

Other religions, notably extreme versions of Islam, promise heaven as the reward for a martyr's death. These ideas are more closely associated with barbarian views of the future than tribal ones: paradise, like Valhalla for the Norse heroes, will be full of the hero's rewards, from perpetually renewed women to ample food and drink and hero's games. But they are also associated, as they were not in the more purely barbarian Norse legends, with a belief in fate, in the will of a god that nothing can avoid or deny. This is the other way for authority to force obedience: the promise of ultimate reward is a very persuasive story.

Barbarians, for whom honour, glory, power, and love, dignity, bravery are the meaningful concepts, get plus points for denying authority and shaping events to their own desires. They have, among their gods and heroes, the mischievous unpredictable ones like Lemminkainen and Puck.

Barbarian nursery stories, like their sagas, laud the hero. They show how luck is associated with particular attitudes, especially a pure heart that does not seek immediate or ultimate reward. There is frequently a test of this purity, from helping a poor blind cripple, who turns out to be a god in disguise, to curing or feeding a desperate animal, who comes to your aid later.

The agents in many of these stories are supernatural – out of the order of things, magical and causeless – 'people', such as fairies (including fairy queens and fairy godmothers), avatars of the gods, demons, and *djinni*. People, especially heroes or aspiring heroes (such as Siegfried, but also Aladdin), attain control over these supernatural beings with the assistance of magic rings, named swords, spells, or merely by their own inner nobility. This changes their fortunes, and luck comes to be on their side; they win battles and bouts against long odds, they climb tall mountains, they kill immortal dragons and monsters. No tribal thinker would even dream of stories like these. For them, fortune favours the well-prepared.

Man is forever inventive, and we have stories that counter even the most heroic tales: the Sidh, the seven-foot-tall elves of *Lords and Ladies* and old Irish folklore, the Devil who buys your soul and has you at his mercy even if you repent, the Grand Vizier, James Bond's opponents.

What is interesting in our discussion of stories here is the characters of these anti-heroes. They don't have any. Elves are the High Folk, but they don't have lives of their own; they are simply portrayed as being antithetic to what people, especially heroes, want to do. We don't care about the human aspects of James Bond's iconic enemies: they are always portrayed as being mindlessly cruel, or avid for power without responsibility and without having to overcome obstacles. They are ciphers, they don't have creative personalities, and they don't learn. If they did, one of them would have shot James Bond dead with a simple gun many years ago, after learning what happens to those who put their trust in laser beams and circular saws. They'd remove his watch first, too.

Rincewind would characterise the elves as 'edge fairies'. They don't tell stories to themselves or, rather, they keep telling the same old story.

It is natural to think of stories as resting on language, but the causality probably works the other way round. Gregory Bateson, in his book *Mind and the Universe*, devotes several chapters to human languages and how we use them to think. But his start on the subject is a beautiful mistake. He starts by looking at an 'outside' view of language, a kind of chemical analogy. Words, he says, are obviously the atoms of language, phrases and sentences the molecules, atoms in combination. Verbs are reactive atoms, link nouns together, and so on. He discusses paragraphs, chapters, books … and fiction, that he claims, very persuasively, is the ultimate triumph of human language.

Bateson shows us a scenario where an audience is watching a murder on stage, and nobody runs to phone the police. And then he goes into another mode, addressing his readers directly. He tells them that he felt that he'd done a really good job on the introduction to language, so he rewarded himself with a visit to the Washington Zoo. Almost the first cage inside the gate had two monkeys playing at fighting, and as he watched them, the whole beautiful edifice that he had written turned upside down in his mind. The monkeys had no verbs, no nouns, no paragraphs. But they understood fiction perfectly.

What does this tell us? Not just that we can rewrite that scene with the boss in our minds. Not even that we can go and see her, and discuss what happened. Its most important implication is that the distinction between fiction and fact sits at the base of language, not at the pinnacle. Verbs and nouns are the most rarefied of abstractions, not the original raw material. We do not acquire stories through language: we acquire language through stories.

FIFTEEN

TROUSER LEG
OF TIME

 IN THE HEAT OF THE NIGHT, magic moved on silent feet.

One horizon was red with the setting sun. This world went around a central star. The elves did not know this. If they had done, it would not have bothered them. They never bothered with detail of that kind. The universe had given rise to life in many strange places, but the elves were not interested in that, either.

This world had created lots of life, too. None of it had ever had what the elves considered to be potential. But this time ...

It had iron, too. The elves hated iron. But this time, the rewards were worth the risk. This time ...

One of them signalled. The prey was close at hand. And now they saw it, clustered in the trees around a clearing, dark blobs against the sunset.

The elves assembled. And then, at a pitch so strange that it entered the brain without the need to use the ears, they began to sing.

'Chmmmmph!' said Archchancellor Ridcully, as a heavy body landed on his back and clamped a hand over his mouth, forcing him back down into the long, dewy grass.

'Listen very carefully!' hissed a voice in his ear. 'When you were small, you had a one-eared toy rabbit called Mr Big Pram! On your sixth birthday your brother hit you on the head with a model boat! And when you were twelve ... do the words "jolly lolly" ring a bell?'

'Mmph!'

'Very well. I'm you. There's been one of those temporal things Mister

160

Stibbons is always goin' on about. I'm taking my hand away now and we'll both quietly crawl away without the elves seeing us. Understand?'

'Mmp.'

'Good man.'

Elsewhere in the bushes the Dean whispered into his own ear: 'Under a secret floorboard in your study—'

Ponder whispered to himself: 'I'm sure we both agree that this should not really be happening ...'

In fact the only wizard who did not bother with concealment was Rincewind, who tapped himself on the shoulder and evinced no surprise at seeing himself. In his life he had seen far more unusual things than his own doppelganger.

'Oh, you,' he said.

''fraid so,' he said glumly.

'Was it you that turned up just now to tell me I should hold my breath?'

'Er ... possibly, but I think I've been superseded by me.'

'Oh. Has Ponder Stibbons being talking about quantum again?'

'You got it in one.'

'Another mess up?'

'More or less. It turns out stopping the elves is a really bad idea.'

'Typical. Do we both survive? There's not much room in the office, what with all the coal—'

'Ponder Stibbons says we may end up remembering everything, because of residual quantum infraction, but we'll sort of be the same person.'

'Any big teeth or sharp edges involved?'

'Not so far.'

'Could be worse, then, all things considered.'

In pairs, the wizards assembled as quietly as they could. Apart from Ridcully, who seemed to quite enjoy his own company, they tried not to look at their doppelgangers; it's quite embarrassing being in the company of someone who knows everything about you, even if that person is yourself.

A few feet away, with the suddenness of lightning, a pale circle appeared on the grass.

'Our transport is here, gentlemen,' said Ponder.

One of the Deans, who was standing well apart from the other Dean, raised his hand.

'What happens to the ones of us that stay behind?' he said.

'It won't matter,' said Ponder Stibbons. 'They'll vanish the moment we do, and the ones of us who end up in the, er, other trouser leg of time will have the memories of both of us. I think that's right, isn't it?'

'Yes,' said Ponder Stibbons. 'A pretty good summation for the layman. So, gentlemen, are we ready? One of everyone, into the circle *now*, please.'

Only the Rincewinds did not move. They knew what to expect.

'Depressing, isn't it,' said one of them, watching the fighting. Both Deans had managed to knock one another out of the circle on the very first charge.

'Especially the way one of the Stibbonses has just laid out the other one with a left hook,' said the other Rincewind. 'An unusual skill in a man of his education.'

'Doesn't give you a lot of confidence, I admit. Toss a coin?'

'Yes, why not …'

They did so.

'Fair enough,' said the winner. 'Nice to have met me.' He picked his way delicately across the groaning bodies and the last couple of struggling wizards, sat down in the centre of the circle of light, and pulled his hat as far down over his head as possible.

A moment later he became, very briefly, a six-dimensional knot and became untied again on a wooden floor in a library.

'Well, that was relatively painful,' he murmured, and looked around.

The Librarian was sitting on his stool. The wizards were around Rincewind, looking amazed and, in some cases, slightly bruised.

Dr Dee was watching them with concern.

'Oh dear, I see it did not work,' he said, and sighed. 'It never works for me, either. I will instruct the servants to fetch some food.'

When he'd gone, the wizards looked at one another.

'Did we *go*?' said the Lecturer in Recent Runes.

'Yes, but we came back at the same time,' said Ponder. He rubbed his chin.

'I can remember *everything*,' said the Archchancellor. 'Amazin'! I was the one that got left behind *and* the one that—'

'Let's just not talk about it, shall we?' said the Dean, brushing his robe.

There was the sound of a muffled voice trying to make itself heard. The Librarian opened his paw.

'Attention please. Attention please,' said Hex.

Ponder took the sphere.

'We're listening.'

'Elves are approaching this property.'

'What, here? In broad daylight?' said Ridcully. 'On our damn world? While we're actually *here*? The nerve!'

Rincewind looked out of the window on to the drive below.

'Is it me,' said the Dean, 'or has it got colder?'

A carriage was rolling up, with a couple of footmen trotting along beside it. It was a fine one, by the standards of the city. There were plumes on the horse. And everything about it was either black or silver.

'It's not just you,' said Rincewind, backing away from the window.

There were sounds at the front door. The wizards heard the distant voice of Dee, and then the creak of the stairs.

'Brethren,' he said, pushing open the door. 'There appears to be a visitor for you downstairs.' He gave them a worried smile. 'A lady ...'

SIXTEEN
FREE WON'T

 WHAT IS THE BIGGEST SOURCE OF DANGER for any organism? Predators? Natural disasters? Fellow organisms of the same species, who constitute the most direct competition for *everything*? Sibling rivals, who compete even in the same family, the same nest? No.

The biggest danger is the future.

If you've survived until now, then your past and present offer no dangers, or at least no new dangers. That time you broke your leg and it didn't heal very well left you vulnerable to lions, but the attack is still going to come, if at all, in the future. You can't do anything to change your past – unless you're a wizard – but you can do something to change your future. In fact, everything you do changes your future, in the sense that the nebulous space of future possibilities starts to crystallise out into the one future that actually happens. If you *are* a wizard, able to visit the past and change that, too, you still have to think about how a range of possibilities crystallises out into just one. You still march forward into your own personal future along your own personal timeline; it's just that, when seen from the perspective of conventional history, that timeline zigzags a lot.

We are committed to a view of ourselves as creatures that exist in time, not just in an ever-changing present. That is why we are fascinated by stories of time travel. And by stories about the future. We have established elaborate methods to foretell the future, and find ourselves at the mercy of deep-seated concepts such as Destiny and Free Will, which relate to our place in time and our ability to change the future – or not. However, we have an ambivalent attitude to the future. In most respects, we think that it is pre-determined, usually by factors beyond our control. Otherwise, how could it be predicted? Most scientific theories of the universe are deterministic: the laws give rise to only one possible future.

To be sure, quantum mechanics involves unavoidable elements of chance, at least according to the orthodox attitude of nearly all physicists, but quantum uncertainty fuzzes out and 'decoheres' as we move from the microscopic world to the macroscopic one, so on a human scale nearly everything that matters is again deterministic from the physical point of view. That doesn't mean that we know ahead of time what's going to happen, though. We have seen that two features of the workings of natural laws, chaos and complexity, imply that deterministic systems need not be predictable in any practical sense. But when we start to think about ourselves, we are utterly certain that we are not deterministic at all. We have free will, we can make choices. We can choose when to get out of bed, what to eat for breakfast, whether or not to put the radio on and listen to the news.

We're not so certain that animals have free will. Do cats and dogs make choices? Or are they merely responding to innate and unchangeable 'drives'? When it comes to simpler organisms like amoebas, we find it difficult to conceive of them choosing between alternatives; though when we watch them through a microscope, we get a strong feeling that they know what they're doing. We're happy to believe that this feeling is an illusion, a silly piece of anthropomorphism, investing human qualities in a tiny bag of biochemicals; no doubt the amoeba is responding, deterministically, to chemical gradients in its environment. But it doesn't look deterministic because of the aforementioned get-outs, chaos and complexity. In contrast, when we make a choice, we have the overwhelming impression that we could have chosen to do something else. If that wasn't possible, then it wasn't really a choice.

We therefore model ourselves as free agents making choice after choice against the background of a complex and chaotic world. We are aware that any threat to our existence – or anything desirable – will come from the future, and that the free choices we make now can and will affect how that future turns out. If only we could foresee the future, we could work out the best choices, and make the future happen the way we would like, and not the way the lions would like. Our intelligence gives us the ability to construct mental models of the future, mostly simple extrapolations of patterns that we have noticed in the past. Our extelligence collects these models, and welds them together into religious prophecies, scientific laws, ideologies, social imperatives … We are time-binding animals, whose every action is constrained not just by the past and present, but also by our own anticipations of the future. We know that we can't predict the future very accurately, but a prediction that works only some of the time is, we feel, better than none. So we tell ourselves and each other stories about the future, and we use those stories to run our lives.

Those stories form part of the extelligence, and they interact with other elements in it, such as science and religion, to create a strong emotional attachment to belief systems or technology that can help us navigate into that uncertain future. Or claim to do so, and can convince us that the claim is valid, even if it's not. In many religions, enormous respect is paid to prophets, people so wise, or so in tune with the deity, that they know what the future will bring. The priests gain respect by predicting eclipses and the turn of the seasons. Scientists gain rather less respect by predicting the movements of the planets, and (less effectively) tomorrow's weather. Whoever controls the future controls human destiny.

Destiny. That's a strange concept in a creature that believes it has free will. If you can control the future, then the future cannot be fixed. If it is not fixed, then there is no such thing as destiny. Unless, perhaps, the future converges on the same events, whatever you do. There are many stories with this theme, of which the most famous is 'Appointment in Samara' (parodied in *The Colour of Magic*), when a man's efforts to escape Death only bring him to the very place where Death is waiting.

We entertain contradictory beliefs about the future. That's not such a surprise: we're not the most logically consistent of creatures. We tend to apply logic locally, within narrow limits, and when it suits us. We're very bad at applying it globally, setting one of our cherished beliefs up against another and looking for the inconsistencies. But we are especially inconsistent when it comes to dealing with the future.

Paradoxically, free will is the last thing you want if you're tribal. You're caught in the matrix of 'Everything that isn't mandatory is forbidden', and there is simply no room for free will. On the one hand, such an existence is very secure; but on the other, punishments and rewards are just as mandatory as everything else if your sins are found out. Your personal responsibility is only to obey the rules.

You can still tell yourself stories about the future, but they involve very narrow choices. 'Shall I attend the ritual meal tonight and leave early to say my evening prayers, or shall I stay for the communal prayers like everybody else?' Even in a tribal system, a lot of cheating goes on, because we're human. 'Well, now … If I leave *early*, then I can drop by Fatima's tent, and my wives won't know about it …'

Plenty of sins are possible, even in a tribal society, and in reality ones that survive allow a little flexibility. If, say, you forget to fast on the Holy Day and someone sees you eating, and you genuinely thought it was tomorrow, or an enemy told you that it was tomorrow, or you had been *made* to think it was tomorrow because an enemy had cast a curse upon you … then some skilful pleading might mitigate your punishment.

The natural and attractive option is always to blame others; it is unbearable to know that you have brought the punishment upon yourself. If you can't see how anyone else can be blamed for material reasons, then blame them for cursing you. Blame Fatima for being attractive and willing, blame an enemy who lied to you. 'Luck' is not available as a concept in a tribal society, because Allah knows everything, Jehovah is omniscient: the natural response is fatalistic acceptance of whatever they throw your way.* If you are to attain heaven, so be it; if your fate is to be flung into the everlasting fires, then that is the Will of God, to which you are subservient. The best

---

* Admittedly, many African tribes think no such thing: you can hide things from the fairly simple local god. But then it's not much of a god. Probably the tribal mores have been corrupted with the passage of time.

you can do, as a peasant-level tribesman, is to find out what is in store for you, what is Written in the Book.

Maybe you don't really want to know what's in the Book, but monkey curiosity overcomes fear, and in any case you can't change what's Written and it might just be nice. So you go to the old lady in the forest who can read tealeaves, or (today) to the iridologist or the spiritualist medium. And all of these alleged ways to foresee the future have a very revealing common feature. They interpret the small-and-contingent into the large-and-important.

Just like the Roman general spilling the guts of a ram on to the ground before the battle, so that the small-and-complicated can mirror a forthcoming battle that will be large-and-complicated, tealeaves and hand-lines are small-and-complicated, and 'must' therefore encode your complicated future. The kind of magic that is being invoked here is an unexpressed homology, which on some level we all believe in because we use it all the time. The stories that we construct in our minds are small-and-complicated, and they really do mirror the large-and-complicated things that happen to us. *The Concise Lexicon of the Occult* lists 93 methods of divination, from aeromancy (divination by the shapes of clouds) to xylomancy (divination by the shapes of twigs). All but four of them employ the small-and-complicated to predict the large-and-complicated; their materials include salt, barley, wind, wax, lead, onion sprouts (that one's called 'cromniomancy'), laughter, blood, fish guts, flames, pearls, and the noises made by mice ('myomancy'). The other four involve invoking spirits, calling up demons, or talking to gods.

To many tribal innocents, other people sometimes seem to have access to different little stories that they can make relevant to your life, like 'Your fate is written on your hand' or 'The dead communicate with me and they know all'. So people of that inclination can convince you, with a bit of flummery, that they know your future, and they can produce convincing large stories which you interpret as your fate.

There is a deep paradox in our attitude to personal free will. We want to know what the future will be in order to make a free choice that protects us against it. So we think of the future of everything outside us as being deterministic, which is why the gypsy or the medium or the dead can know what it is going to be. Nevertheless, we think of our own future as involving free choices. Our free will lets us choose to consult the gypsy, who then convinces us that we have no choice: for example, that the life-line on our palm determines when we will die. So our actions betray a deep-seated belief that the laws of the universe apply to everything except us.

The biggest wholesale business that preys on our convictions and confusions about free will in a powerful and often cruel universe, is astrology. Astrologers claim authority from Ancient Egypt, from Paracelsus and Dee, from Ancient Wisdom of all kinds including the Hindu Vedas and other Eastern literature. Let us review the appeal of astrology in the light of narrativium.

Astrologers have an immense following, and they have managed to pick up on both tribal and barbarian stories. They have the counterscientific story for the civilised culture, able to attract both the tribal and barbarian aspects of our foolishness. They really do believe that the future, for each of us, is influenced by our time of birth.* They time it to the second.

What seems to be important to them is against which starry background (the Zodiac) we view the planets in our own solar system. As we move from intra-uterine life to the hands of the midwife, doctor, partner, our lives are determined from then on by astral forces. This strange belief is supported by so many people, who turn to the 'Your Stars' pages first in their daily newspaper, that we should seek some explanation within our 'story' framework. What is the story of our futures that is implicit in the control of our lives by the positions of the stars? As opposed, say, to the medical staff who, at the time of our

* Why *birth*, the sheerest accident during our development? Why not fertilisation? Or hatching from the *zona pellucida*, the egg membrane? Or the first heartbeat? Or the first dream (while still in the uterus)? Or the first word, or the first carnal experience? There *are* aspects of our future that are determined by, at least, the date of our birth (we may end up the youngest or the oldest child in the school intake that year, and that can make a big difference) but we're not talking about these human-created things here.

birth, probably had more gravitational influence upon us* than the planet Jupiter was having?

Well, the stars are obviously very numinous, powerful. They're up there wheeling over us. At least, they were when we were shepherds, staying out all night, but most civilised folk now don't know why the Moon changes its shape, let alone why or where the pole star is. Yes, all right, *you* do, and it's not surprising. Others don't, and don't think what they don't know is worth knowing.

They have a vague feel for a few of the constellations, especially the Big Dipper (or Great Bear), but they don't know that those stars are not near to each other, but merely appear to be in that formation when viewed from Earth, and then only for a short time, astronomically speaking. Most people don't entertain astronomical thoughts, so why are the stars so heavily involved in our most potent stories? Perhaps because, in our nursery stories, the celestial sphere gives a context, a primitive animistic one in which Moon and Sun take protagonist parts? We don't find that persuasive. Perhaps it is because the power of the stars entered our cultural stories back in the time when everyone could see the clear night sky, and has hung on. Or perhaps it is the jargon of the Zodiac-mongers, with their gypsy fortune-teller use of language to give received certainty to the most nebulous of prophecies. We've never heard anyone say, after reading the newspaper's astrology columns, 'Right, then, they're totally wrong today, no more astrology for me!'

There are others playing the same card, from Pyramidologists to Ancient Astronaut promoters to Flying Saucers Will Save Us visionaries to Rosicrucians. Regular UFO enthusiasts and Loch Ness monster photographers are much less dangerous. We focus on the prophets: those who, like followers of Nostradamus's prophecies or astrology, must believe that all the little contingencies add up to a grand pattern of the human future and that Fate rules us all.

This is the tribal interpretation of the feeling of free will: it is an

---

* The gravitational attraction exerted by a single doctor at a distance of 6 inches is roughly twice that of Jupiter at its closest point to the Earth.

illusion, for God already knows our futures. Kismet (the word comes from the Turkish 'qismet' and Arabic 'qisma') rules. Moreover – a neat twist that gives power over people as well as their money – whether you will be a beetle or a king in the next turn of the cosmic wheel is determined by the balance that you have achieved in this life. This is equally out of your control, in practice, but you can escape to an inner life, making it as far as possible irrelevant to the vicissitudes that attack your outer self, and thereby avoid beetlehood in your next incarnation.

That apparent escape again depends on our ability to construct stories about our future. Here, our future divides, with the soul taking one direction under our own control and freed from the control of powerful others, while the body is manifestly bowed by slavery, starvation, or torture. Hundreds of millions have found comfort in that apparent control of their futures, following the story of their spiritual selves and denying the pains of the material self.

In the Buddhist literature and practice, something close to that transcendence seems to be achievable. If you believe in fate, or the nearby concept of karma, then wisdom can consist only in foreseeing events, training your spiritual self to accept what happens, and teaching others to do the same. Some authority will provide your map of material events, but your destiny cannot be avoided by fighting it. Your only option is to lead a disciplined spiritual life, guided by stories of previous successes in this quest, notably the Buddha, and to entertain hopes of leaving the Wheel of Life altogether, to exist as a spiritual presence with all ties to the material severed.

This nirvanic view of heaven is not for those who enjoy the material ride too much to want to get off the bus. And the paradoxical nature of the prophetic predictions – of all prophetic predictions – is disturbing. There is no way at all that a deterministic Earth can be accommodated by today's view of what planets are like, and most of today's more sophisticated religions have no room for an immanent God, tinkering with each life, and its context, to achieve its destiny. Those that do have room for immanence encounter real problems with modern technology, whose basis lies in ways-of-the-universe modelled by science, not by djinns or the whim of a deity or deities. And although we may, with Fredric Brown, be amused that when the *djinni* that worked the electric light and the radio came out on strike, the steam-power genies came out in sympathy, we enjoy this animistic fantasy as fuel for Murphy's Law and nice Disneyesque animations. We don't buy any of it for real causality.

Joseph Needham brought light to this kind of confusion. He pointed out, in the introduction to his truly gigantic *History of Science in China*, that the reason why China never developed science as the West knows it is that they never espoused monotheism. In polytheistic philosophies, it isn't very sensible to search for *the* cause of something, like a thunderstorm, say: you're liable to get a very contingent answer involving several incidents in the love lives of the gods, and an explanation of the provenance of thunderbolts that verges on the ridiculous.*

Monotheists, however, by which we mean someone like Abraham, to whom we shall return later, reckon that God had a consistent set of ideas and causalities in mind when he set the universe up. *One* set of ideas. If you expect your one God to be consistent, then it's worth asking how those causalities relate to each other: for example, 'black clouds and rain will be associated with thunderstorms when ...' whatever. The monotheist can predict the weather, even if rather badly. But the polytheist needs a theopsychologist and a precise account of what the gods are up to at the moment. She needs to know whether a tiff between two gods will result in a thunderstorm. So scientific causality is compatible with God-causality, but not with gods-causality.

Monotheists, moreover, have a built-in intolerance. The position that there is only one truth, only one avenue to the one God, sets each monotheistic religion in opposition to all others. There is no room for manoeuvre, no way to tolerate the manifest errors of people who believe in some other god. So monotheism laid the foundation for the Inquisition, and for intemperate Christianity through the ages from the crusades through to African and Polynesian missionaries. 'I have the story, and it is the only one' is characteristic of many cults, all of them intolerant.

---

* At least on Discworld you can *see* the gods acting disgracefully.

Faiths, of course, do get along. But they get along because of the hammering they have taken at the hands of science, material development and better education. They get along because of wise people within them who recognise the commonality of humanity. Where there are too few wise people, you get Northern Ireland. If you are lucky.

If the future is not fixed, but malleable, and we can predict the effects of our present behaviour, however badly, then predicting the future can be self-defeating. And that can even be the reason for predicting it.

Most of the Biblical prophets seem, like many science-fiction authors today, to be warning against what might happen if we go on as we are doing. So they succeed when their prophecy is *not* correct, because people heed it and change their actions. We can understand that; even though the prophecy didn't come true, we can all see that it might have done: it has given us a better idea of the phase space that the future of our culture lives in.

What about the gypsy who prophesies that a tall dark man will come into your life, thus making you receptive to all those future tall dark men? (if tall dark men interest you, of course; it's up to you.) This could be a self-fulfilling prophecy, the opposite of the stories told by Biblical prophets. It's a story that the recipient is sympathetic to, wants to happen.

There are said to be only seven basic story plots, so perhaps our minds are much less varied than we think, so that the newspaper astrologer and the fortune-teller are navigating a much smaller phase space of human experience than we thought. This would account for so many people feeling that the predictions show deep insight.

But when *astronomers* predict the future, *and get it right*, people are, paradoxically, much less impressed. When they predict eclipses correctly, every time, this seems less meaningful than the astrologers nearly getting many people right, sometimes. Remember Y2K, the prophecy that planes would fall out of the sky soon after the year 2000 dawned and your toaster wouldn't work? That prophecy cost the world several billion dollars in work to avert the problem – and it didn't happen. A waste of time, then? Not at all. It didn't happen because people took precautions. If they hadn't done, the cost would have been much higher. It was a Biblical prophecy: 'If this goes on ...' And, lo, the multitude heeded.

This recursive dependence of prophecy upon people's responses to it, unlike most of the other kinds of thing that we say, relates back to our facility with our own made-up little futures, the stories that we tell ourselves. They confirm us in our identities. It is no wonder that when someone – an astrologer or Nostradamus, say – pokes his finger into this mental place where we live, and inserts some of his own stories, we want to believe him. His stories are more exciting than ours. We wouldn't have thought, going down the stairs to get a train to work, 'I wonder if I'm going to meet a tall dark guy today?' But once it's been put into our minds, we smile at all the dark men, even some quite short ones. And so our lives are changed (perhaps in quite major ways, if you are a man doing the smiling) as are the stories that we ourselves proposed for our futures.

This way that we react, fairly predictably, to what the world throws at us, casts doubt on our otherwise unshakeable belief that *we get to choose what we do*. Do we truly possess free will? Or are we like the amoeba, drifting this way and that, propelled by the dynamic of a phase space that cannot be perceived from outside?

In *Figments of Reality* we included a chapter with the title 'We wanted to have a chapter on free will but we decided not to, so here it is'. There we examined such issues as whether, in a world without genuine free will, it would be fair to blame a person for their actions. We conclude that in a world without genuine free will, there might not be any choice: they would get blamed anyway because the possibility of them not being blamed did not exist.

We won't go over that ground in detail, but we do want to summarise the main thrust of the argument. We start by observing that there is no effective scientific test for free will. You can't run the universe again, with everything *exactly* as it was, and see if a different choice can be made second time round. Moreover, there seems to be no room in the laws of physics for genuine free will. Quantum indeterminacy, seized on so readily by many philosophers and scientists as a catch-all explanation of 'consciousness', is the wrong kind of thing altogether: random unpredictability is not the same as choosing between clear alternatives.

There are many ways in which the known laws of physics could offer an *illusion* of free will, for example by exploiting chaos or emergence, but there is no way to set up a system that could make different choices even though every particle in the universe, including those making up the system, is in the same state on both occasions.

Add to this one rather interesting aspect of human social behaviour: although we feel as if *we* have free will, we don't act as if we believe that anybody else has. When somebody does something uncharacteristic, 'not like them', we don't say 'Oh, Fred is exercising his free will. He's been a lot happier since he smiled at the tall, dark stranger.' We say 'What the devil has got into Fred?' Only when we find a reason for his actions, an explanation *not* involving the exercise of free will (like drunkenness, or 'doing it for a bet') do we feel satisfied.

All of this suggests that our minds do not actually make choices: they make judgements. Those judgements reveal not what we have chosen, but what kind of mind we possess. 'Well, I never would have guessed,' we say, and feel we've learned something that we can use in future dealings with that person.

So what about that strong feeling that we get, of making a choice? That's not what we're doing, it's what it *feels like to us* when we're doing it, just as that vivid grey quale of the visual system is not actually out there on the elephant, but an added decoration that exists in our heads. 'Choosing' is what our minds feel like from inside when they're judging between alternatives. Free will is not a real attribute of human beings at all: it is merely the quale of judgement.

SEVENTEEN

FREEDOM OF INFORMATION

 PEOPLE BELIEVED THAT ELVES could look like anything they wanted to, but this was, strictly speaking, wrong. Elves looked the same all the time (rather dull and grey, with large eyes, rather similar to bushbabies without the charm) but they could, without effort, cause others to see them differently.

Currently the Queen looked like a fashionable lady of the time, in black lace, sparkling here and there with diamonds. Only with a hand over one eye and extreme concentration could even a trained wizard dimly see the true nature of the elf, and even then his eye would water alarmingly.

Nevertheless, the wizards stood up when she entered. There's such a thing as courtesy, after all.

'Welcome to my world, gentlemen,' said the Queen, sitting down. Behind her, a couple of guards took up station either side of the door.

'Ours!' snarled the Dean. 'It's *our* world!'

'Let us continue to disagree, shall we?' said the Queen brightly. 'You may have constructed it, but it's our world now.'

'We have iron, you know,' said Ridcully. 'Would you like some tea, by the way? Foul stuff made without actual tea.'

'Much good may it do you. No, thank you,' said the Queen. 'Please note that my guards are human. So is your host. The Dean looks angry. You intend to *fight* here? When you have no magic? Be serious, gentlemen. You should be grateful, after all. This is a world without narrativium. Your strange humans were monkeys without stories. They did not know how the world was supposed to *go*. We gave them stories, and made them people.'

'You gave them gods and monsters,' said Ridcully. 'Stuff that stops people thinking straight. Superstitions. Demons. Unicorns. Bogeymen.'

'You have bogeymen on your world, don't you?' said the Queen.

'Yes, we do. But outside, where we can get at 'em. They ain't stories. When you can *see* 'em, they don't have any power.'

'Like unicorns,' said the Lecturer in Recent Runes. 'When you *meet* one, you find out it's just a big sweaty horse. Looks nice, smells horsey.'

'And it's *magical*,' said the Queen, her eyes gleaming.

'Yes, but that's just another thing about it,' said Ridcully. 'Big, sweaty, magical. There's nothing *mysterious* about it. You just learn the rules.'

'But surely you should be pleased!' said the Queen, her eyes saying that she knew they weren't and was glad of it. 'Everyone here thinks this world is just like yours! Many people even believe that it is flat!'

'Yes, but back home they'd be right,' said Ridcully. 'Here they're just ignorant.'

'Well, there is not a single thing you can do about it,' said the Queen. 'This is *our* world, Mister Wizard. It's all stories. The religions here ... amazing! And the beliefs ... wonderful! The crop is bountiful, the harvest is rewarding. Do you know that more people believe in magic here than they do on your world?'

'*We* don't have to believe in it. It *works!*' snapped Ridcully.

'They believe in it here, and it *doesn't*,' said the Queen. 'And thus they believe in it even more, while ceasing to believe in themselves. Isn't it astonishing?'

She stood up. Most of the wizards went to stand up, too, and one or two of them got all the way. Misogynists to a man, the wizards were therefore always punctiliously polite to ladies.

'Here, you are just rather fussy old men,' she said. 'But *we* understand this world and we have had time to cultivate it. We like it. You can't take us away. Your humans need us. We are part of their world now.'

'This world, madam, has about another thousand years before all life is wiped out,' said Ridcully.

'Then there are other worlds,' said the Queen, lightly.

'That's *all* you have to say?'

'What else is there? Worlds begin and end,' said the Queen. 'That is how the universe works. That is the great circle of existence.'

'The great circle of existence, madam, can eat my underwear!' said Ridcully.

'Fine words,' said the Queen. 'You are good at concealing your true thoughts from me, but I can also see them in your face, nevertheless. You think you can still fight us and win. You have forgotten that there is *no* narrativium in this world. It does not know how stories should go. *Here*, the third son of a king is probably just a useless weak prince. *Here*, there are no heroes, only degrees of villainy. An old lady gathering wood in the forest is just an old lady and not, as in your world, almost certainly a witch. Oh, there is a *belief* in witches. But a witch here is merely a method of ridding society of burdensome old ladies and an inexpensive way of keeping the fire going all night. Here, gentlemen, good does not ultimately triumph at the expense of a few bruises and a non-threatening shoulder wound. Here, evil is generally defeated by a more organised kind of evil. My world, gentlemen. Not yours. Good day to you.'

And then she was gone.

The wizards sat down again. Outside, the carriage rattled away.

'Quite well spoken for an elf, I thought,' said the Lecturer in Recent Runes. 'Good turn of phrase.'

'And that's *it*?' said Ridcully. 'We *can't* do anything?'

'We don't have any magic, sir,' said Ponder.

'But we *do* know everythin' is goin' to turn out all right, though, don't we?' said Ridcully. 'We know that people get off the planet before the next big wallop, right? We saw the evidence. Right?'

Ponder sighed.

'Yes, sir. But it might not happen. It's like the Shell Midden people.'

'They *didn't* happen?'

'Not … here, sir,' said Ponder.

'Ah. And you're goin' to say "it's because of quantum" at some point?'

'I hadn't intended to, sir, but you're on the right lines.'

'So ... when we left them, did they pop out of existence?'

'No, sir. We did.'

'Oh. Well, so long as *someone* did ...' said Ridcully. 'Any thoughts, gentlemen?'

'We could go to the pub again?' said the Lecturer in Recent Runes, hopefully.

'No,' said Ridcully. 'This is serious.'

'So am I.'

'I don't see what we can do,' said the Dean. 'The humans here needed the elves to tinker with their heads. When we stopped that, we got the Shell Midden people. When we *didn't* stop it, we got people like Dee, head half full of rubbish.'

'I know someone who'd be right at home with this problem,' said Ridcully, thoughtfully. 'Mister Stibbons, we would be able to get back home now, wouldn't we? Just to send a semaphore message?'

'Yes, sir, but there's no need for that. Hex can do that directly,' said Ponder, before he could stop himself.

'How?' said Ridcully.

'I ... er ... connected him up to the semaphore just after you left, sir. Er ... it was just a matter of pulleys and things. Er ... I installed a little set of repeater arms on the roof of the High Energy Magic Building. Er ... and employed a gargoyle to do the watching, and we needed one anyway, because the pigeons up there have really got too numerous ... er ...'

'So Hex can send and receive messages?' said Ridcully.

'Yes, sir. All the time. Er ...'

'But that costs a fortune! Is it coming out of your budget, man?'

'Er, no, sir, because it's actually quite cheap, er ... it's free, actually ...' Ponder went for broke. 'Hex worked out the codes, you see. The gargoyles up on the big tower don't bother about where the signals are coming from, they just notice the codes, so, er, Hex started by adding the codes for the Assassins' Guild or the Fools' Guilds to the messages and, er, they probably didn't notice the extra amount on their bills because they're using the clacks all the time these days—'

'So ... we're *stealing*?' said Ridcully.

'Well, er, yes, sir, in a way, but it's hard to know exactly *what*. Last month Hex worked out the semaphore company's own codes so his messages travel as part of their internal signalling, sir. No one gets billed for that.'

'This is very disturbing news, Stibbons,' said Ridcully sternly.

'Yes, sir,' said Ponder, looking at his feet.

'I feel I must ask you a rather difficult and worrying question: is it likely that anyone will find out?'

'Oh, no, sir. It's impossible to trace.'

'Impossible?'

'Yes, sir. Every week Hex sends a message to company headquarters readjusting the total of messages sent, sir. Anyway, there's so many I don't think anyone checks.'

'Oh? Well, that's all right then,' said Ridcully. 'It never really happens, and no one can find out it's us in any case. Can we send all our messages that way?'

'Well, technically yes, sir, but I think that might be abusing the—'

'We *are* academics, Stibbons,' said the Dean. 'And information should be allowed to flow freely.'

'Exactly,' said the Lecturer in Recent Runes. 'An untrammelled flow of information is essential to a progressive society. This is the age of the semaphore, after all.'

'Obviously it flows *to* us,' said Ridcully.

'Oh, certainly,' said the Dean. 'We don't want it flowing *away* from us. We're talking about flow here, not *spread*.'

'You wanted a message sent?' said Ponder, before the wizards got too deeply into this.

'And we really don't have to pay?' said Ridcully.

Ponder sighed. 'No, sir.'

'Jolly good,' said the Archchancellor. 'Have this one sent to the kingdom of Lancre, will you? They've only got one clacks tower. Got your notebook? Message begins:

"To Mistress Esmerelda Weatherwax. How are you? I am fine. An interesting problem has arisen ..."

EIGHTEEN

BIT FROM IT

 A SEMAPHORE IS A SIMPLE and time-honoured example of a digital communication system. It encodes letters of the alphabet using the positions of flags, lights, or something similar. In 1795 George Murray invented a version that is close to the system currently used in Discworld: a set of six shutters that could be opened or closed, thus giving 64 different 'codes', more than enough for the entire alphabet, numbers 0 to 10 and some 'special' codes. The system was further developed but ceased to be cutting-edge technology when the electric telegraph heralded the wired age. The Discworld semaphore (or 'clacks') has been taken *much* further, with mighty trunk route towers carrying bank after bank of shutters, aided by lamps after dark, and streaming messages bi-directionally across the continent. It is a pretty accurate 'evolution' of the technology: if we too had failed to harness steam and electricity, we might well be using something like it ...

There is enough capacity on that system even to handle pictures – seriously. Convert the picture to a 64 × 64 grid of little squares that can be black, white or four shades of grey, and then read the grid from left to right and top to bottom like a book. It's just a matter of information, a few clever clerks to work out some compression algorithms, and a man with a shallow box holding 4,096 wooden blocks, their six sides being, yes, black, white and four shades of grey. It'll take them a while to reassemble the pictures, but clerks are cheap.

Digital messages are the backbone of the Information Age, which is the name we currently give to the one we're living in, in the belief that we know a lot more than anyone else, ever. Discworld is comparably proud of being in the Semaphore Age, the Age of the Clacks. But what, exactly, is information?

When you send a message, you are normally expected to pay for it – because if you don't, then whoever is doing the work of transmitting that message for you will object. It is this feature of messages that has got Ridcully worried, since he is wedded to the idea that academics travel free.

Cost is one way to measure things, but it depends on complicated market forces. What, for example, if there's a sale on? The scientific concept of 'information' is a measure of how much message you're sending. In human affairs, it seems to be a fairly universal principle that for any given medium, longer messages cost more than short ones. At the back of the human mind, then, lurks a deep-seated belief that messages can be quantified: they have a *size*. The size of a message tells you 'how much information' it contains.

Is 'information' the same as 'story'? No. A story does convey information, but that's probably the least interesting thing about stories. Most information doesn't constitute a story. Think of a telephone directory: lots of information, strong cast, but a bit weak on narrative. What counts in a story is its meaning. And that's a very different concept from information.

We are proud that we live in the Information Age. We do, and that's the trouble. If we ever get to the Meaning Age, we'll finally understand where we went wrong.

Information is not a thing, but a concept. However, the human tendency to reify concepts into things has led many scientists to treat information as if it is genuinely real. And some physicists are starting to wonder whether the universe, too, might be made from information.

How did this viewpoint come about, and how sensible is it?

Humanity acquired the ability to quantify information in 1948, when the mathematician-turned-engineer Claude Shannon found a way to define how much information is contained in a message – he preferred the term *signal* – sent from a transmitter to a receiver using some kind of code. By a signal, Shannon meant a series of binary digits ('bits', 0 and 1) of the kind that is ubiquitous in modern computers and communication devices, and in Murray's semaphore. By a code, he meant a specific procedure that transforms an original signal into another one. The simplest code is the trivial 'leave it alone'; more sophisticated codes can be used to detect or even correct transmission errors. In the engineering applications, codes are a central issue, but for our purposes here we can ignore them and assume the message is sent 'in plain'.

Shannon's information measure puts a number to the extent to which our uncertainty about the bits that make up a signal is reduced by what we receive. In the simplest case, where the message is a string of 0s and 1s and every choice is equally likely, the amount of information in a message is entirely straightforward: it is the total number of binary digits. Each digit that we receive reduces our uncertainty about that *particular* digit (is it 0 or 1?) to certainty ('it's a 1', say) but tells us nothing about the others, so we have received one bit of information. Do this a thousand times and we have received a thousand bits of information. Easy.

The point of view here is that of a communications engineer, and the unstated assumption is that we are interested in the bit-by-bit content of the signal, not in its meaning. So the message 111111111111111 contains 15 bits of information, and so does the message 111001101101011. But Shannon's concept of information is not the only possible one. More recently, Gregory Chaitin has pointed out that you can quantify the extent to which a signal contains *patterns*. The way to do this is to focus not on the size of the message, but on the size of a computer program, or *algorithm*, that can generate it. For instance, the first of the above messages can be created by the algorithm 'every digit is a 1'. But there is no simple way to describe the second message, other than to write it down bit by bit. So these two messages have the same Shannon information content, but from Chaitin's point of view the second contains far more 'algorithmic information' than the first.

Another way to say this is that Chaitin's concept focuses on the extent to which the message is 'compressible'. If a short program can generate a long message, then we can transmit the program instead of the message and save time and money. Such a program 'compresses' the message. When your computer takes a big graphics file – a photograph, say – and turns it into a much smaller file in JPEG format, it has used a standard algorithm to compress the information in the original file. This is possible because photographs contain numerous patterns: lots of repetitions of blue pixels for the sky, for instance. The more incompressible a signal is, the more information in Chaitin's sense it contains. And the way to compress a signal is to describe the patterns that make it up. This implies that incompressible signals are random, have no pattern, yet contain the most information. In one way this is reasonable: when each successive bit is maximally unpredictable, you learn more from knowing what it is. If the signal reads 111111111111111 then there is no great surprise if the next bit turns out to be 1; but if the signal reads 111001101101011 (which we obtained by tossing a coin 15 times) then there is no obvious guess for the next bit.

Both measures of information are useful in the design of electronic technology. Shannon information governs the time it takes to transmit a signal somewhere else; Chaitin information tells you whether there's a clever way to compress the signal first, and transmit something smaller. At least, it would do if you could calculate it, but one of the features of Chaitin's theory is that it is impossible to calculate the amount of algorithmic information in a message – and he can prove it. The wizards would approve of this twist.

'Information' is therefore a useful concept, but it is curious that 'To be or not to be' contains the same Shannon information as, and *less* Chaitin information than, 'xyQGRlfryu&d%sk0wc'. The reason for this disparity is that information is not the same thing as meaning. That's fascinating. What really matters to people is the meaning of a message, not its bit-count, but mathematicians have been unable to quantify meaning. So far.

And that brings us back to stories, which are messages that convey meaning. The moral is that we should not confuse a story with 'information'. The elves gave humanity stories, but they didn't give them any information. In fact, the stories people came up with included things like werewolves, which don't even *exist* on Roundworld. No information there – at least, apart from what it might tell you about the human imagination.

Most people, scientists in particular, are happiest with a concept when they can put a number to it. Anything else, they feel, is too vague to be useful. 'Information' is a number, so that comfortable feeling of precision slips in without anyone noticing that it might be spurious. Two sciences that have gone a long way down this slippery path are biology and physics.

The discovery of the 'linear' molecular structure of DNA has given evolutionary biology a seductive metaphor for the complexity of organisms and how they evolve, namely: *the genome of an organism represents the information that is required to construct it*. The origin of this metaphor is Francis Crick and James Watson's epic discovery that an organism's DNA consists of 'code words' in the four molecular 'letters' A C T G, which, you'll recall, are the initials of the four possible 'bases'. This description led to the inevitable metaphor that the genome contains information about the corresponding organism. Indeed, the genome is widely described as 'containing the information needed to produce' an organism.

The easy target here is the word 'the'. There are innumerable reasons why a developing organism's DNA does not determine the organism. These non-genomic influences on development are collectively known as 'epigenetics', and they range from subtle chemical tagging of DNA to the investment of parental care. The hard target is 'information'. Certainly, the genome includes information in some sense: currently an enormous international effort is being devoted to listing that information for the human genome, and also for other organisms such as rice, yeast, and the nematode worm *Caenorhabditis elegans*. But notice how easily we slip into cavalier attitudes, for here the word 'information' refers to the human mind as receiver, not to the developing organism. The Human Genome Project informs *us*, not organisms.

This flawed metaphor leads to the equally flawed conclusion that the genome explains the complexity of an organism in terms of the amount of information in its DNA code. Humans are complicated because they have a long genome that carries a lot of information; nematodes are less complicated because their genome is shorter. However, this seductive idea can't be true. For example, the Shannon information content of the human genome is smaller by several orders of magnitude than the quantity of information needed to describe the wiring of the neurons in the human brain. How can we be more complex than the information that describes us? And some amoebas have much longer genomes than ours, which takes us down several pegs as well as casting even more doubt on DNA as information.

Underlying the widespread belief that DNA complexity explains organism complexity (even though it clearly doesn't) are two assumptions, two scientific stories that we tell ourselves. The first story is *DNA as Blueprint*, in which the genome is represented not just as an important source of control and guidance over biological development, but as the information needed to determine an organism. The second is *DNA as Message*, the 'Book of Life' metaphor.

Both stories oversimplify a beautifully complex interactive system. *DNA as Blueprint* says that the genome is a molecular 'map' of an organism. *DNA as Message* says that an organism can pass that map to the next generation by 'sending' the appropriate information.

Both of these are wrong, although they're quite good science fiction – or, at least, interestingly bad science fiction with good special effects.

If there is a 'receiver' for the DNA 'message' it is not the next generation of the organism, which does not even exist at the time the 'message' is being 'sent', but the ribosome, which is the molecular machine that turns DNA sequences (in a protein-coding gene) into protein. The ribosome is an essential part of the coding system; it functions as an 'adapter', changing the sequence information along the DNA into an amino acid sequence in proteins. Every cell contains many ribosomes: we say 'the' because they are all identical. The metaphor of DNA as information has become almost universal, yet virtually nobody has suggested that the ribosome must be a vast repository of information. The structure of the ribosome is now known in high detail, and there is no sign of obvious 'information-bearing' structure like that in DNA. The ribosome seems to be a fixed 'machine'. So where has the information gone? Nowhere. That's the wrong question.

The root of these misunderstandings lies in a lack of attention to context. Science is very strong on content, but it has a habit of ignoring 'external' constraints on the systems being studied. Context is an important but neglected feature of information. It is so easy to focus on the combinatorial clarity of the message and to ignore the messy, complicated processes carried out by the receiver when it decodes the message. Context is crucial to the interpretation of messages: to their meaning. In his book *The User Illusion* Tor Nørretranders introduced the term *exformation* to capture the role of the context, and Douglas Hofstadter made the same general point in *Gödel, Escher, Bach*. Observe how, in the next chapter, the otherwise incomprehensible message 'THEOSTRY' becomes obvious when context is taken into account.

Instead of thinking about a DNA 'blueprint' encoding an organism, it's easier to think of a CD encoding music. Biological development is like a CD that contains instructions for building a new CD-player. You can't 'read' those instructions without already having one. If meaning does not depend upon context, then the code on the CD should have an *invariant* meaning, one that is independent of the player. Does it, though?

Compare two extremes: a 'standard' player that maps the digital code on the CD to music in the manner intended by the design engineers, and a jukebox. With a normal jukebox, the only message that you send is some money and a button-push; yet in the context of the jukebox these are interpreted as a specific several minutes' worth of music. In principle, any numerical code can 'mean' any piece of music you wish; it just depends on how the jukebox is set up, that is, on the exformation associated with the jukebox's design. Now consider a jukebox that reacts to a CD not by playing the tune that's encoded on it, as a series of bits, but by interpreting that code as a number, and then playing some other CD to which that number has been assigned. For instance, suppose that a recording of Beethoven's Fifth Symphony starts, in digital form, with 11001. That's the number 25 in binary. So the jukebox reads the CD as '25', and looks for CD number 25, which we'll assume is a recording of Charlie Parker playing jazz. On the other hand, elsewhere in the jukebox is CD number 973, which actually is Beethoven's Fifth Symphony. Then a CD of Beethoven's Fifth can be 'read' in two totally different ways: as a 'pointer' to Charlie Parker, or as Beethoven's Fifth Symphony itself (triggered by whichever CDs start with 973 in binary). Two contexts, two interpretations, two meanings, two results.

Whether something *is* a message depends upon context, too: sender and receiver must agree upon a protocol for turning meanings into symbols and back again. Without this protocol a semaphore is just a few bits of wood that flap about. Tree branches are bits of wood that flap about, too, but no one ever tries to decode the message being transmitted by a tree. Tree rings – the growth rings that appear when you saw through the trunk, one ring per year – are a different matter. We have learned to 'decode' their 'message', about climate in the year 1066 and the like. A thick ring indicates a good year with lots of growth on the tree, probably warm and wet; a thin ring indicates a poor year, probably cold and dry. But the sequence of tree rings only became a message, only conveyed information, when we figured out the rules that link climate to tree growth. The tree didn't *send* its message to us.

In biological development the protocol that gives meaning to the DNA message is the laws of physics and chemistry. That is where the exformation resides. However, it is unlikely that exformation can be quantified. An organism's complexity is not determined by the *number* of bases in its DNA sequence, but by the complexity of the actions initiated by those bases within the context of biological development. That is, by the *meaning* of the DNA 'message' when it is received by a finely tuned, up-and-running biochemical machine. This is where we gain an edge over those amoebas. Starting with an embryo that develops little flaps, and making a baby with those exquisite little hands, involves a series of processes that produce skeleton, muscles, skin, and so on. Each stage depends on the current state of the others, and all of them depend on contextual physical, biological, chemical and cultural processes.

A central concept in Shannon's information theory is something that he called *entropy*, which in this context is a measure of how statistical patterns in a source of messages affect the amount of information that the messages can convey. If certain patterns of bits are more likely than others, then their presence conveys *less* information, because the uncertainty is reduced by a smaller amount. In English, for example, the letter 'E' is much more common than the letter 'Q'. So receiving an 'E' tells you less than receiving a 'Q'. Given a choice between 'E' and 'Q', your best bet is that you're going to receive an 'E'. And you learn the most when your expectations are proved wrong. Shannon's entropy smooths out these statistical biases and provides a 'fair' measure of information content.

In retrospect, it was a pity that he used the name 'entropy', because there is a longstanding concept in physics with the same name, normally interpreted as 'disorder'. Its opposite, 'order', is usually identified with complexity. The context here is the branch of physics known as thermodynamics, which is a specific simplified model of a gas. In thermodynamics, the molecules of a gas are modelled as 'hard spheres', tiny billiard balls. Occasionally balls collide, and when they do, they bounce off each other as if they are perfectly elastic. The Laws of Thermodynamics state that a large collection of such spheres will obey certain statistical regularities. In such a system, there are two forms of energy: mechanical energy and heat energy. The First Law states that the total energy of the system never changes. Heat energy can be transformed into mechanical energy, as it is in, say, a steam engine; conversely, mechanical energy can be transformed into heat. But the sum of the two is always the same. The Second Law states, in more precise terms (which we explain in a moment), that heat cannot be transferred from a cool body to a hotter one. And the Third Law states that there is a specific temperature below which the gas cannot go – 'absolute zero', which is around -273 degrees Celsius.

The most difficult – and the most interesting – of these laws is the Second. In more detail, it involves a quantity that is again called 'entropy', which is usually interpreted as 'disorder'. If the gas in a room is concentrated in one corner, for instance, this is a more ordered (that is, less disordered!) state than one in which it is distributed uniformly throughout the room. So when the gas is uniformly distributed, its entropy is higher than when it is all in one corner. One formulation of the Second Law is that the amount of entropy in the universe always increases as time passes. Another way to say this is that the universe always becomes less ordered, or equivalently less complex, as time passes. According to this interpretation, the highly complex world of living creatures will inevitably become less complex, until the universe eventually runs out of steam and turns into a thin, lukewarm soup.

This property gives rise to one explanation for the 'arrow of time', the curious fact that it is easy to scramble an egg but impossible to unscramble one. Time flows in the direction of increasing entropy. So scrambling an egg makes the egg more disordered – that is, increases its entropy – which is in accordance with the Second Law. Unscrambling the egg makes it less disordered, and decreases energy, which conflicts with the Second Law. An egg is not a gas, mind you, but thermodynamics can be extended to solids and liquids, too.

At this point we encounter one of the big paradoxes of physics, a source of considerable confusion for a century or so. A different set of physical laws, Newton's laws of motion, predicts that scrambling an egg and unscrambling it are equally plausible physical events. More precisely, if *any* dynamic behaviour that is consistent with Newton's laws is run backwards in time, then the result is also consistent with Newton's laws. In short, Newton's laws are 'time-reversible'.

However, a thermodynamic gas is really just a mechanical system built from lots of tiny spheres. In this model, heat energy is just a special type of mechanical energy, in which the spheres vibrate but do not move *en masse*. So we can compare Newton's laws with the laws of thermodynamics. The First Law of Thermodynamics is simply a restatement of energy conservation in Newtonian mechanics, so the First Law does not contradict Newton's laws. Neither does the Third Law: absolute zero is just the temperature at which the spheres cease vibrating. The amount of vibration can never be less than zero.

Unfortunately, the Second Law of Thermodynamics behaves very differently. It *contradicts* Newton's laws. Specifically, it contradicts the property of time-reversibility. Our universe has a definite direction for

its 'arrow of time', but a universe obeying Newton's laws has two distinct arrows of time, one the opposite of the other. In our universe,
scrambling eggs is easy and unscrambling them seems impossible.
Therefore, according to Newton's laws, in a time-reversal of our universe, unscrambling eggs is easy but scrambling them is impossible.
But Newton's laws are the same in both universes, so they cannot prescribe a definite arrow of time.

Many suggestions have been made to resolve this discrepancy. The
best mathematical one is that thermodynamics is an approximation,
involving a 'coarse-graining' of the universe in which details on very
fine scales are smeared out and ignored. In effect, the universe is
divided into tiny boxes, each containing (say) several thousand gas
molecules. The detailed motion inside such a box is ignored, and only
the average state of its molecules is considered.

It's a bit like a picture on a computer screen. If you look at it from
a distance, you can see cows and trees and all kinds of structure. But
if you look sufficiently closely at a tree, all you see is one uniformly
green square, or pixel. A real tree would still have detailed structure
at this scale – leaves and twigs, say – but in the picture all this detail
is smeared out into the same shade of green.

In this approximation, once 'order' has disappeared below the level
of the coarse-graining, it can never come back. Once a pixel has been
smeared, you can't unsmear it. In the real universe, though, it sometimes can, because in the real universe the detailed motion inside the
boxes is still going on, and a smeared-out average ignores that detail.
So the model and the reality are *different*. Moreover, this modelling
assumption treats forward and backward time asymmetrically. In forward time, once a molecule goes into a box, it can't escape. In
contrast, in a time-reversal of this model it can escape from a box but
it can never get in if it wasn't already inside that box to begin with.

This explanation makes it clear that the Second Law of
Thermodynamics is not a genuine property of the universe, but merely
a property of an approximate mathematical description. Whether the
approximation is helpful or not thus depends on the context in which
it is invoked, not on the content of the Second Law of Thermodynamics.
And the approximation involved destroys any relation with Newton's

laws, which are inextricably linked to that fine detail.

Now, as we said, Shannon used the same word 'entropy' for his measure of the structure introduced by statistical patterns in an information source. He did so because the mathematical formula for Shannon's entropy looks exactly the same as the formula for the thermodynamic concept. Except for a minus sign. So thermodynamic entropy looks like negative Shannon entropy: that is, thermodynamic entropy can be interpreted as 'missing information'. Many papers and books have been written exploiting this relationship – attributing the arrow of time to a gradual loss of information from the universe, for instance. After all, when you replace all that fine detail inside a box by a smeared-out average, you lose information about the fine detail. And once it's lost, you can't get it back. Bingo: time flows in the direction of information-loss.

However, the proposed relationship here is bogus. Yes, the formulas look the same ... but they apply in very different, unrelated, contexts. In Einstein's famous formula relating mass and energy, the symbol c represents the speed of light. In Pythagoras's Theorem, the same letter represents one side of a right triangle. The letters are the same, but nobody expects to get sensible conclusions by identifying one side of a right triangle with the speed of light. The alleged relationship between thermodynamic entropy and negative information isn't quite that silly, of course. Not *quite*.

As we've said, science is not a fixed body of 'facts', and there are disagreements. The relation between Shannon's entropy and thermodynamic entropy is one of them. Whether it is meaningful to view thermodynamic entropy as negative information has been a controversial issue for many years. The scientific disagreements rumble on, even today, and published, peer-reviewed papers by competent scientists flatly contradict each other.

What seems to have happened here is a confusion between a formal mathematical setting in which 'laws' of information and entropy can be stated, a series of physical intuitions about heuristic interpretations of those concepts, and a failure to understand the role of context. Much is made of the resemblance between the formulas for entropy in information theory and thermodynamics, but little attention is paid to the context in which those formulas apply. This habit has led to some very sloppy thinking about some important issues in physics.

One important difference is that in thermodynamics, entropy is a quantity associated with a *state* of the gas, whereas in information theory it is defined for an information *source*: a system that generates entire collections of states ('messages'). Roughly speaking, a source is a phase space for successive bits of a message, and a message is a trajectory, a path, in that phase space. In contrast, a thermodynamic configuration of molecules is a *point* in phase space. A specific configuration of gas molecules has a thermodynamic entropy, but a specific message does not have a Shannon entropy. This fact alone should serve as a warning. And even in information theory, the information 'in' a message is not negative information-theoretic entropy. Indeed the entropy of the source remains unchanged, no matter how many messages it generates.

There is another puzzle associated with entropy in our universe. Astronomical observations do not fit well with the Second Law. On cosmological scales, our universe seems to have become more complex with the passage of time, not less complex. The matter in the universe started out in the Big Bang with a very smooth distribution, and has become more and more clumpy – more and more complex – with the passage of time. The entropy of the universe seems to have decreased considerably, not increased. Matter is now segregated on a huge range of scales: into rocks, asteroids, planets, stars, galaxies, galactic clusters, galactic superclusters and so on. Using the same metaphor as in thermodynamics, the distribution of matter in the universe seems to be becoming increasingly ordered. This is puzzling since the Second Law tells us that a thermodynamic system should become increasingly disordered.

The cause of this clumping seems to be well established: it is gravity. A second time-reversibility paradox now rears its head. Einstein's field equations for gravitational systems are time-reversible. This means that if any solution of Einstein's field equations is time-reversed, it becomes an equally valid solution. Our own universe, run backwards in this manner, becomes a gravitational system that gets less and less

clumpy as time passes – so getting less clumpy is just as valid, physically, as getting more clumpy. Our universe, though, does only one
of these things: more clumpy.

Paul Davies's view here is that 'as with all arrows of time, there is
a puzzle about where the asymmetry comes in … The asymmetry must
therefore be traced to initial conditions'. What he means here is that
even with time-reversible laws, you can get different behaviour by
starting the system in a different way. If you start with an egg and stir
it with a fork, then it scrambles. If you start with the scrambled egg,
and very very carefully give each tiny particle of egg exactly the right
push along precisely the opposite trajectory, then *it will unscramble*.
The difference lies entirely in the initial state, not in the laws. Notice
that 'stir with a fork' is a very general kind of initial condition: lots of
different ways to stir will scramble the egg. In contrast, the initial condition for unscrambling an egg is extremely delicate and special.

In a way this is an attractive option. Our clumping universe is like
an unscrambling egg: its increasing complexity is a consequence of
very special initial conditions. Most 'ordinary' initial conditions would
lead to a universe that isn't clumped – just as any reasonable kind of
stirring leads to a scrambled egg. And observations strongly suggest
that the universe's initial conditions at the time of the Big Bang were
extremely smooth, whereas any 'ordinary' state of a gravitational system presumably should be clumped. So, in agreement with the
suggestion just outlined, it seems that the initial conditions of the universe must have been very special – an attractive proposition for those
who believe that our universe is highly unusual, and ditto for our
place within it.

From the Second Law to God in one easy step.

Roger Penrose has even quantified how special this initial state is,
by comparing the thermodynamic entropy of the initial state with that
of a hypothetical but plausible final state in which the universe has
become a system of Black Holes. This final state shows an extreme
degree of clumpiness – though not the ultimate degree, which would
be a single giant Black Hole. The result is that the entropy of the
initial state is about 10^{-30} times that of the hypothetical final state, making it extremely special. So special, in fact, that Penrose was led to

introduce a new time-asymmetric law that forces the early universe to be exceptionally smooth.

Oh, how our stories mislead us … There is another, much more reasonable, explanation. The key point is simple: gravitation is very different from thermodynamics. In a gas of buzzing molecules, the uniform state – equal density everywhere – is stable. Confine all the gas into one small part of a room, let it go, and within a split second it's back to a uniform state. Gravity is exactly the opposite: uniform systems of gravitating bodies are unstable. Differences smaller than any specific level of coarse-graining not only *can* 'bubble up' into macroscopic differences as time passes, but do.

Here lies the big difference between gravity and thermodynamics. The thermodynamic model that best fits our universe is one in which differences dissipate by disappearing below the level of coarse-graining as time marches forwards. The gravitic model that best fits our universe is one in which differences amplify by bubbling up from below the level of coarse-graining as time marches forwards. The relation of these two scientific domains to coarse-graining is exactly opposite when the same arrow of time is used for both.

We can now give a completely different, and far more reasonable, explanation for the 'entropy gap' between the early and late universes, as observed by Penrose and credited by him to astonishingly unlikely initial conditions. It is actually an artefact of coarse-graining. Gravitational clumping bubbles up from a level of coarse-graining to which thermodynamic entropy is, by definition, insensitive. Therefore virtually *any* initial distribution of matter in the universe would lead to clumping. There's no need for something extraordinarily special.

The physical differences between gravitating systems and thermodynamic ones are straightforward: gravity is a long-range attractive force, whereas elastic collisions are short-range and repulsive. With such different force laws, it is hardly surprising that the behaviour should be so different. As an extreme case, imagine systems where 'gravity' is so short range that it has no effect unless particles collide, but then they stick together forever. Increasing clumpiness is obvious for such a force law.

The real universe is both gravitational and thermodynamic. In some contexts, the thermodynamic model is more appropriate and thermodynamics provides a good model. In other contexts, a gravitational model is more appropriate. There are yet other contexts: molecular chemistry involves different types of forces again. It is a mistake to shoehorn all natural phenomena into the thermodynamic approximation or the gravitic approximation. It is especially dubious to expect both thermodynamic and gravitic approximations to work in the same context, when the way they respond to coarse-graining is diametrically opposite.

See? It's simple. Not magical at all ...

Perhaps it's a good idea to sum up our thinking here.

The 'laws' of thermodynamics, especially the celebrated Second Law, are statistically valid models of nature in a particular set of contexts. They are *not* universally valid truths about the universe, as the clumping of gravity demonstrates. It even seems plausible that a suitable measure of gravitational complexity, like thermodynamic entropy but different, might one day be defined – call it 'gravtropy', say. Then we might be able to deduce, mathematically, a 'second law of gravitics', stating that the gravtropy of a gravitic system *increases* with time. For example, gravtropy might perhaps be the fractal dimension ('degree of intricacy') of the system.

Even though coarse-graining works in opposite ways for these two types of system, both 'second laws' – thermodynamic and gravitic – would correspond rather well to our own universe. The reason is that both laws are formulated to correspond to what we actually observe in our own universe. Nevertheless, despite this apparent concurrence, the two laws would apply to drastically different physical systems: one to gases, the other to systems of particles moving under gravity.

With these two examples of the misuse of information-theoretic and associated thermodynamic principles behind us, we can turn to the intriguing suggestion that the universe is made from information.

Ridcully suspected that Ponder Stibbons would invoke 'quantum' to explain anything really bizarre, like the disappearance of the Shell Midden People. The quantum world *is* bizarre, and this kind of invocation is always tempting. In an attempt to make sense of the quantum

universe, several physicists have suggested founding all quantum phenomena (that is, everything) on the concept of information. John
Archibald Wheeler coined the phrase 'It from Bit' to capture this idea.
Briefly, every quantum object is characterised by a finite number of
states. The spin of an electron, for instance, can either be up or down,
a binary choice. The state of the universe is therefore a huge list of
ups and downs and more sophisticated quantities of the same general
kind: a very long binary message.

So far, this is a clever and (it turns out) useful way to formalise the
mathematics of the quantum world. The next step is more controversial. All that really matters is that message, that list of bits. And what
is a message? Information. Conclusion: the *real* stuff of the universe
is raw information. Everything else is made from it according to quantum principles. Ponder would approve.

Information thereby takes its place in a small pantheon of similar
concepts – velocity, energy, momentum – that have made the transition from convenient mathematical fiction to reality. Physicists like to
convert their technically most useful mathematical concepts into real
things: like Discworld, they reify the abstract. It does no physical harm
to 'project' the mathematics back into the universe like this, but it may
do philosophical harm if you take the result literally. Thanks to a similar process, for example, entirely sane physicists today insist that our
universe is merely one of trillions that coexist in a quantum super-
position. In one of them you left your house this morning and were
hit by a meteorite; in the one in which you're reading this book, that
didn't happen. 'Oh, yes,' they urge: 'those other universes *really do
exist*. We can do experiments to prove it.'

Not so.

Consistency with an experimental result is not a proof, not even a
demonstration, that an explanation is valid. The 'many-worlds' concept, as it is called, is an interpretation of the experiments, within its
own framework. But any experiment has many interpretations, not all
of which can be 'how the universe really does it'. For example, all
experiments can be interpreted as 'God made that happen', but those
selfsame physicists would reject their experiment as a proof of the existence of God. In that they are correct: it's just one interpretation. But

then, so are a trillion coexisting universes.

Quantum states do superpose. Quantum universes can also superpose. But separating them out into classical worlds in which real-life people do real-life things, and saying that *those* superpose, is nonsense. There isn't a quantum physicist anywhere in the world that can write down the quantum-mechanical description of a person. How, then, can they claim that their experiment (usually done with a couple of electrons or photons) 'proves' that an alternate you was hit by a meteorite in another universe?

'Information' began its existence as a human construct, a concept that described certain processes in communication. This was 'bit from it', the abstraction of a metaphor from reality, rather than 'it from bit', the reconstruction of reality from the metaphor. The metaphor of information has since been extended far beyond its original bounds, often unwisely. Reifying information into the basic substance of the universe is probably even more unwise. Mathematically, it probably does no harm, but Reification Can Damage Your Philosophy.

NINETEEN

LETTER FROM LANCRE

GRANNY WEATHERWAX, KNOWN TO ALL and not least to herself as Discworld's most competent witch, was gathering wood in the forests of Lancre, high in the mountains and far from any university at all.

Wood gathering was a task fraught with danger for an old lady so attractive to narrativium. It was quite hard these days, when gathering firewood, to avoid third sons of kings, young swineherds seeking their destiny and others whose unfolding adventure demanded that they be kind to an old lady who would *with a certainty* turn out to be a witch, thus proving that smug virtue is its own reward.

There is only a limited number of times even a kindly disposed person wishes to be carried across a stream that they had, in fact, not particularly desired to cross. These days, she kept a pocket full of small stones and pine cones to discourage that kind of thing.

She heard the soft sound of hooves behind her and turned with a pine cone raised.

'I *warn* you, I'm fed up with you lads always on the ear'ole for three wishes—' she began.

Shawn Ogg, astride his official donkey, waved his hands desperately.*

---

* Lancre was so backward that its population of 500 had only one civil servant, Shawn Ogg, who handled everything from national defence and tax gathering to mowing the castle lawns, although he was allowed help with the lawns. Lawn required care.

'It's me, Mistress Weatherwax! I wish you'd stop doing this!'

'See?' said Granny. 'You ain't havin' another two!'

'No, no, I've just come up to deliver this for you ...'

Shawn waved quite a thick wad of paper.

'What is it?'

''Tis a clacks for you, Mistress Weatherwax! It's only the third one we've ever had!' Shawn beamed at the thought of being so close to the cutting edge of technology.

'What's one of them things?' Granny demanded.

'It's like a letter that's taken to bits and sent through the air,' said Sean.

'By them towers I keep flyin' into?'

'That's right, Mistress Weatherwax.'

'They move 'em around at night, you know,' said Granny. She took the paper.

'Er ... I don't think they do ...' Shawn ventured.

'Oh, so I don't know how to fly a broomstick right, do I?' said Granny, her eyes glinting.

'Actually, yes, I've remembered,' said Shawn quickly. 'They move them around *all the time*. On carts. Big, *big* carts. They ...'

'Yes, yes,' said Granny, sitting on a stump. 'Be quiet now, I'm readin' ...'

The forest went silent, except for the occasional shuffling of paper.

Finally, Granny Weatherwax finished. She sniffed. Birdsong came back into the forest.

'Silly old fools think they can't see the wood for the trees, and the trees *are* the wood,' she muttered. 'Cost a lot, does it, sendin' messages like this?'

'That message,' said Shawn, in awe, 'cost more than 600 dollars! I counted the words! Wizards must be made of money!'

'Well, I ain't,' said the witch. 'How much is one word?'

'Five pence for the sending and five pence the first word,' said Shawn, promptly.

nny. She frowned in concentration, and her lips moved

ver been one for numbers,' she said, 'but I reckon that

pence and one half-penny?'

Sean knew his witches. It was best to give in right at the start.

'That's right,' he said.

'You have a pencil?' said Granny. Shawn handed it over. With great care, the witch printed some block capitals on the back of one of the pages, and gave it to him.

'That's all?' he said.

'Long question, short answer,' said Granny, as it if was some universal truth. 'Was there anything else?'

Well, there might be the money, Sean thought. But in her own localised way, Granny Weatherwax had an academic position in these matters. Witches took the view that they helped society in all kinds of ways which couldn't easily be explained but would become obvious if they stopped doing them, and that it was worth six pence and one half-penny not to find out what these were.

He didn't get his pencil back.

The hole into L-space was quite obvious now. It fascinated Dr Dee, who was confidently expecting angels to come out of it, although all it had produced so far was an ape.

The wizards' automatic response to any problem was to see if there was a book about it. L-space was providing plenty of books. The difficulty, however, was finding the ones that applied to the current history; when you potentially know everything, it's hard to find anything you want to know.

'So let's see where we are now, shall we?' said Ridcully, after a while. 'The last known books in this leg of the trousers of time are due to be written in—?'

'About a hundred years' time,' said the Lecturer in Recent Runes, looking at his notes. 'Just before the collapse of civilisation, such as it is. Then there's fire, famine, war ... all the usual stuff.'

'Hex says people here are back to living in villages when the asteroid hits,' said Ponder. 'Things are rather better on one or two other continents, but no one even sees it coming.'

'There have been periods like it before,' said the Dean. 'But as far as we can tell, in the area where we are now there were always small isolated groups of people who preserved what books there were.'

'Ah. Our kind of people,' said Ridcully.

'Afraid not,' said the Dean. 'Religious.'

'Oh dear,' said Ridcully.

'It's hard to follow, but there appear to be about four main gods on this continent,' said the Dean. 'Loosely associated.'

'Big beards in the sky?' said Ridcully.

'A couple, yes.'

'Clearly a morphic memory of ourselves, then,' said Ridcully.

'It's hard to tell, with religions,' said the Dean. 'But at least they preserved the idea that books were important and that reading and writing were more than just a skive for people too weedy to hack at one another with swords.'

'Any of these religious places still around?' said the Lecturer in Recent Runes. 'Would it be useful to drop in, explain that we are, in fact, the creators of this universe, and put them right on a few points?'

There was silence. And then Ponder said, in his best talking-to-superiors voice: 'I believe, sir, that this world is no different from our own in its attitude to apparent human beings who turn up and say that they're a god.'

'We wouldn't get special treatment?'

'Not of the sort you have in mind, sir, no,' said Ponder. 'Besides, the places in this country appear to have been closed down by a recent monarch. I'm not sure I understand it all, but it appears to have been some kind of cost-cutting exercise.'

'Downsizing of redundant units, re-allocation of staff, that sort of thing?' said Ridcully.

'Yes, sir,' said Ponder. 'And a few murders, some torturing, that sort of thing.'

'But probably nothing, I'm sure, that couldn't be sorted out by getting everyone to go and run around in the woods shooting paint at one another,' said the Lecturer in Recent Runes, innocently.

'I shall ignore that, Runes,' said Ridcully. 'Now, gentlemen, we *are* supposed to be thinkers. We haven't got magic. We *can* be moved in time and space, according to Hex. And we've got big sticks. What *can* we do?'

'A message has arrived,' said Hex.

'From Lancre? That's quick!'

'Yes. The message is unsigned. It is: THEOSTRY.'

Hex spelled it out. Ponder wrote it down in his notebook.

'What does that mean?' Ridcully looked up at his wizards.

'Looks a bit religious to me,' said the Dean. 'Rincewind? This sort of thing is right up your street, isn't it?'

Rincewind looked at the word. Really, when you came to think of it, his whole *life* was a crossword puzzle ...

'The clacks people charge by the word, don't they?' he said.

'Yes, it's scandalous,' said Ridcully. 'Five pence a word, on the long-distance trunk!'

'And this was sent back by an old woman in Lancre, where as far as I recall the chicken is the basic unit of currency?' said Rincewind. 'Not much money for fancy messages, then. It looks to me like a simple anagram of ... THE STORY ...'

'I think it means "change the story",' said Ponder, without looking up. 'At a saving of five pence.'

'We *tried* changing it!' said the Dean.

'Change it in a different way, perhaps? At a different time?' said Ponder. 'We've got L-space. We ought to be able to get some guidance from books written in different futures—'

'Ook!'

'I'm sorry, sir, but the library rules don't apply here!' said Ponder.

'Look at it this way, old chap,' said Ridcully, to the angry Librarian. 'The rules do of course apply here, everyone can see that, and we wouldn't dream of asking you to interfere with the nature of causality in the normal way of things. It's just that the nature of causality on this world is such that, if any libraries survive the next thousand years without being used for lighting fires or uncomfortable toilet paper, they're due to be destroyed in a fireball and/or entombed in ice. Dr Dee's wonderful books which you like so much, with their many delicate illustrations of completely useless magical circles and rather interesting mathematical cyphers, will go the way of the, the ...' He snapped his fingers. 'Someone give me the name of something that'll be going completely extinct,' he demanded.

'People,' said Rincewind.

There was silence.

Then the Librarian said: 'Ook ook.'

'He says he's just *finding* the books, okay?' said Rincewind. 'And he'll leave them in a pile and go out of the room and no one is to look at them while he's gone, because if they do *he won't know about it*, and if he coughs loudly before he comes back in that will only be because he's got a cough *and not for any other reason*, okay?'

TWENTY
SMALL GODS

 'Religious,' said the Dean.

'Oh dear,' said Ridcully.

Discworld's wizardry is not terribly keen on religion. Given the history of the Discworld, this is not surprising. One big problem is that on Discworld, gods are *known* to be real. We list a few later on, but we can set the scene with reference to the god of mayflies. In *Reaper Man*, an old mayfly is telling some youngsters about this god, as they hover just above the surface of a stream:

'... you were telling us about the Great Trout.'

'Ah. Yes. Right. The Trout. Well, you see, if you've been a good mayfly, zigzagging up and down properly—'

'—taking heed of your elders and betters—'

'— yes, and taking heed of your elders and betters, then eventually the Great Trout—'

Clop.

Clop.

'Yes?' said one of the younger mayflies.

There was no reply.

'The Great Trout what?' said another mayfly, nervously.

They looked down at a series of expanding concentric rings on the water.

'The holy sign!' said a mayfly. 'I remember being told about that! A Great Circle in the water! Thus shall be the sign of the Great Trout!'

Roundworld religions avoid the difficulty of gods that you can actually see, or meet or be eaten by: most of the world's current religions find it best to go the whole hog and locate their gods in a place that is not just outside Roundworld the planet, but outside Roundworld the universe. This demonstrates admirable foresight, for regions impenetrable today may be a forest of tourist hotels tomorrow. When the sky was an unexplored and unfathomable realm, it was fashionable to locate gods in the sky, or on top of unscalable Mount Olympus, or in the halls of Valhalla, which amounts to much the same thing. But now all significant mountains have been climbed, people routinely fly across the Atlantic, five miles up, and reports of encounters with gods are few.

However, it turns out that when gods don't manifest themselves in physical form on an everyday basis, they acquire an impressive degree of ineffability. On Discworld, on the other hand, it is possible to run into gods in the street or even in the gutter. They also lounge around in Discworld's equivalent of Valhalla, known as Dunmanifestin, which is situated on top of Cori Celesti, a ten mile high spire of green ice and grey stone at the Disc's hub.

Because of the everyday presence of tangible gods, on Discworld there's no problem about *belief* in gods; it's more a matter of how much you disapprove of their lifestyle. On Roundworld, deities do not infest the highways and byways – or, if they do, they do so in such a subtle guise that the unbeliever does not notice them. It then becomes possible to have a serious debate about belief, because that's what most people's concept of God rests on.

We've already said that on Discworld everything is reified, and that's pretty much the case there with belief. Now B-space, the space of beliefs, is huge, because people have vivid and varied imaginations and can believe almost anything. Therefore G-space, the space of gods, is also huge. And on Discworld, phase spaces are reified. So the Discworld not only *has* gods: it is infested with them. There are at least 3,000 major gods on the Disc, and scarcely a week passes without the research theologians discovering more. Some use props like false noses to appear in religious chronicles under hundreds of different names, which makes it difficult to keep count accurately. Among them are Cephut, the god of cutlery (*Pyramids*), Flatulus, god of the winds (*Small Gods*), Grune, the god of unseasonal fruit (*Reaper Man*), Hat, the vulture-headed god of unexpected guests (*Pyramids*), Offler, the crocodile god (*Mort* and *Sourcery*), Petulia, the goddess of negotiable affection (*Small Gods*), and Steikheigel, the god of isolated cow byres (*Mort*).

Then there are the minor gods. According to *The Discworld Companion*, 'There are billions of them, tiny bundles containing nothing more than a pinch of pure ego and some hunger'. What they hunger for, at least to start with, is human belief, because on Discworld the size and power of a god is proportional to how many people believe in him, her, or it. Things are much the same on Roundworld, in fact, because the influence and power of a religion are proportional to the number of its adherents. So the parallel is much closer than you might expect – which is what you should always expect with Discworld, because it has an uncanny ability to reflect and illuminate the human condition in Roundworld. Actually, it's not always human (or mayfly) belief that matters. According to *Lords and Ladies*:

There were a number of gods in the mountains and forests of Lancre. One of them was known as Herne the Hunted. He was a god of the chase and the hunt. More or less.

Most gods are created and sustained by belief and hope. Hunters danced in animal skins and created gods of the chase, who tended to be hearty and boisterous with the tact of a tidal wave. But they are not the only gods of hunting. The prey has an occult voice too, as the blood pounds and the hounds bay. Herne was the god of the chased and the hunted and all small animals whose ultimate destiny is to be an abrupt damp squeak.

When discussing religious beliefs, there is always the danger of upsetting people. The same goes when discussing football, of course, but people take their religion nearly as seriously. So let us begin by acknowledging, as we did towards the end of *The Science of Discworld*, that 'all religions are true, for a given value of true'. We have no wish to damage your beliefs, if you have them, or to damage your lack of beliefs if you don't. We don't mind if we cause you to modify your beliefs, though. That's your responsibility and your choice: don't blame us. But we're shortly going to have a go at science, and then we're going to have a go at art, so we don't think it's fair that religion should get away scot-free. Anyway, whatever your beliefs, religion is an essential feature of the human condition, and it's one of the things that made us what we are. We *have* to examine it, and ask whether Discworld puts it in a new light.

If you are religious, and you want to feel comfortable about what we're saying, you can always assume that we're talking about all the *other* religions, but not yours. Some years ago, during Ecumenical Week, Rabbi Lionel Blue was giving the 'Thought for the Day' on BBC Radio 4, as part of a series on tolerance. He was the first speaker in the series, and he ended with a joke. 'They shouldn't have asked me to start the series,' he said, and then explained how the later speakers from other religions would differ from him, and how he would be tolerant about that. 'After all,' he said, 'they worship God in their way ... whereas I worship Him in His.'

If you see that this *is* a joke, as the good rabbi did, but also understand that outside that cosy context this is not, in a multicultural world, a good way to think, let alone speak, then you're already getting to grips with the ambivalent role that religion has played in human history. And with the mental twists and turns required to live in a multiculture.

The big problem with religion, for a dispassionate observer, has nothing to do with belief versus proof. If religion were susceptible to scientific-style proofs or disproofs, there wouldn't be a lot to argue about. No, the big problem is the disparity between individual human spirituality – the deep-seated feeling that we *belong* in this awesome universe – with the unmitigated disasters that organised, large-scale religions have at various times, including in all probability yesterday, inflicted on the planet and its people. This is upsetting. Religion ought to be a force for good, and mostly it is ... But when it isn't, it goes spectacularly and horribly wrong.

In both *Pyramids* and *Small Gods*, we see that the real problem in this connection is not religion as such, but priests. Priests have been known to seize upon the spiritual feelings of individuals and twist them into something terrible; the Quisition in *Small Gods* was hardly an invention. Sometimes it had been done for power, or for money. It's even been done because the priests really believe that this is what the god of choice wants them to do.

Again, on an individual level many priests (or equivalent) are perfectly nice people who do many positive things, but collectively they can have some very negative effects. It is this mismatch that will form the core of our discussion, because it tells us interesting things about what it is to be human.

We are very tiny, fragile creatures inside a huge, uncontrollable universe. Evolution has equipped us not just with eyes to see the universe, but minds to hold little models of it within us; that is, to tell ourselves stories about it.

We have learned, over the millennia, to exert more and more control over our world, but we see evidence every day that our ability to control our own lives is extremely limited. In the past, disease, death, famine and ferocious animals were part of everyday existence. You could control when you planted your crops, but you couldn't control when the rains came, and you might just get jumped by a pride of lionesses while you were bending down to pull up weeds.

It is very uncomfortable to have to cope unaided with that kind of world, and many people still have to do so. Everyone feels much happier if they *believe* that there are ways to control rain and lionesses.

Now, the human mind is an inveterate pattern-seeker, one that finds patterns even where none exist. Every week millions of perfectly sane people look for patterns in lottery numbers, oblivious to the absence of any meaningful structure in random numbers. So it's not really necessary for the belief in an ability to control rain or lionesses to correspond to an actual ability to do so. We all know that even when things are under control, they can still go wrong, so our faith in our beliefs seldom gets seriously challenged, whatever happens.

The idea that there is a Rain Goddess who decides when it will rain, or a Lion God who can either keep you safe from lion attacks or unleash them upon you, therefore has irresistible advantages. You can't control rain, and of course you can't control a Rain Goddess either, but, with the proper rituals, you can hope to influence her decisions. This is where the priesthood comes in, because they can act as an intermediary between everybody else and the gods. They can prescribe the appropriate rituals – and, like all good politicians, they can claim the credit when things work out and blame someone else when they go wrong. 'What, Henry was eaten by a lion? Well then, he must not have shown proper respect when making his daily sacrifice to the Lion God.' 'How do you know that?' 'Well, if he had shown proper respect, he wouldn't have been eaten.' Ally that to the priests' soon-acquired power to throw you to the earthly representatives of the Lion God if you disagree, and you can see that the Cult of the Lion God has an awful lot going for it.

People look at the universe around them, and they feel overawed. It's so big, so incomprehensible – yet it seems to dance to a tune. People who grow up in a culture – especially one with a lengthy history and a well-developed set of techniques for making buildings, planting crops, hunting animals, building boats – immediately recognise that they are faced with *something* that is far greater than they are. Which immediately raises all the big philosophical questions: where did it come from, what's it for, why am I here? And so on.

Imagine how it must have seemed to Abraham, one of the founding fathers of Judaism. He was probably a shepherd, and he probably lived in and around Ur, one of the first true city-states. He was surrounded by the icons of simple-minded religions: gold-plated idols, masks, altars. He was wildly unimpressed by them. They were trivial things, small-minded. They did not begin to measure up to the awesomeness of the natural world, and its stunning power. Additionally, he was aware that 'something' much bigger than him was running that world. It knew when to plant crops and when to reap them, how to tell whether rain was on the way, how to build boats, how to breed sheep (well, he would have known that bit), how to have a prosperous life. Even more: it knew how to pass all this knowledge on to the next generation. Abraham knew that his own tiny intelligence was nothing compared to this majestic *something*. So he reified it, and gave it a name: Jehovah, which means 'that which is'. So far, so good, but then he made a simple but intellectually fatal error. He fell for the trap of 'ontic dumping'.

Nice phrase. What does it mean? Ontology is the study of knowledge. Not knowledge itself, just its study. One important way to firm up new knowledge is to invent new words. For instance, when you make an arrow, someone has to produce the sharp pointy thing that sits at its business end. They chip it from flint or cast it in bronze; either way, you can't go on forever referring to it as 'the sharp pointy thing on the end of an arrow'. So you cast around for a metaphor, and you remember that the thing that sits at the business end of a person or animal is called its head. So you invent the term 'arrow-head'.

You have now *dumped* the knowledge of what the flint or bronze gadget is into a name. We say 'dumped', because for most purposes you don't need to recall where the name came from. Arrowhead (no hyphen) has now become a thing in its own right, not a property possessed in relation to an arrow.

The human mind is a storytelling device, a metaphor machine: ontic dumping comes naturally to creatures like us. It's how our language works, how our minds work. It's a trick we use to simplify things that would otherwise be incomprehensible. It is the linguistic analogue of a political hierarchy as a way for one person to control millions. As a side effect, ontically dumped words wallow in associations. We are seldom conscious of these, except when we occasionally stop and ask something like 'What on Earth does "gossamer" mean?' Then we rush off to the dictionary and discover that it probably (no one ever knows these things for sure) comes from 'goose summer'. What's that got to do with fine threads that float on the breeze? Well, in a summer when geese abound, a *good* summer, you find a lot of these fine spider-silk threads hanging in the air …

Subconsciously, though, we are all too aware of the dark associations several layers down in the ontic-dumping hierarchy. So words, which ought to be abstract labels, are smeared all over with their own (often irrelevant) stories.

Abraham, then, was overawed by 'that which is', and he ontically dumped it into a word, Jehovah. Which quickly became a thing, indeed, a *person*. That's another of our habits, personifying things. So Abraham made the tiny step from 'there is something outside us that is greater than ourselves' to 'there is some*one* outside us who is greater than ourselves'. He had looked on the burgeoning extelligence of his own culture, and before his eyes it turned into God.

And that made so much sense. It explained so much else. Instead of the world being like it was for reasons he couldn't understand – even though that greater something clearly understood it perfectly well – he now saw that the world was like that because God had made it that way. The rain fell not because some tawdry idol rain-god made it fall; Abraham was too smart to believe *that*. It fell because that awesome God whose presence could be seen everywhere made it fall. And he, Abraham, couldn't hope to understand the Mind of God, so of course he couldn't predict when it would rain.

We have used Abraham here as a placeholder. Choose your religion, choose your founder, adapt the story to fit. We're not saying that we know that the birth of Judaism happened the way we've just explained. That was just a story, probably no more true than Winnie-the-Pooh and the honey. But just as Pooh in the rabbit-hole teaches us about greed, so Abraham's ontic dumping points to a plausible route whereby sane, sensitive people can be led from their own private spiritual feelings to reify a natural process into an unfathomable Being.

This reification has had many positive consequences. People take notice of the wishes of unfathomable, all-powerful Beings. Religious teachings often lay down guidelines (laws, commandments) for acceptable behaviour towards other people. To be sure, there are many disagreements between the different religions, or between sects within a given religion, about points of fine detail. And there are some quite substantial areas of disagreement, such as the recommended treatment of women, or to what extent basic rights should be extended to the infidel. On the whole, however, there is a strong consensus in such teachings, for example an almost universal condemnation of theft and murder. Virtually all religions reinforce a very similar consensus of what constitutes 'good' behaviour, perhaps because it is this consensus that has survived the test of time. In terms of the barbarian/tribal distinction, it is a tribal consensus, reinforced by tribal methods such as ritual, but none the worse for that.

Many people find inspiration in their religion, and it helps instil a sense of belonging. It enhances their feeling of what an awesome place the universe is. It helps them cope with disasters. With exceptions, mainly related to specific circumstances such as war, most religions preach that love is good and hatred is bad. And throughout history, ordinary people have made huge sacrifices, often of their own lives, on that basis.

This kind of behaviour, generally referred to as altruism, has caused evolutionary biologists a great deal of head-scratching. First, we'll summarise how they have thought about the problem and what kinds of conclusion they have reached. Then we'll consider an alternative approach, originally motivated by religious considerations, which looks to us to be far more promising.

At first sight, altruism is not a problem. If two organisms cooperate, by which in this context we here mean that each is willing to risk its life to help the other,* then both stand to gain. Natural selection favours such an advantage, and reinforces it. What more explanation is needed?

Quite a lot, unfortunately. A standard reflex in evolutionary biology is to ask whether such a situation is stable – whether it will persist if some organisms adopt other strategies. What happens, for example, if most organisms cooperate, but a few decide to cheat? If the cheats prosper, then it is better to become a cheat than to cooperate, and the strategy of cooperation is unstable and will die out. Using the methods of mid-twentieth-century genetics, the approach pioneered by Ronald Aylmer Fisher, you can do the sums and work out the circumstances in which altruism is an evolutionarily stable strategy. The answer is that it all depends upon whom you cooperate with, whose life you risk your own to save. The closer kin they are to you, the more genes they share with you, so the more worthwhile it is for you to risk your own safety. This analysis leads to conclusions like 'It is worth jumping into a lake to save your sister, but not to save your aunt.' And certainly not to save a stranger.

That's the genetic orthodoxy, and like most orthodoxies, it is believed by the orthodox. On the other hand, though: if someone has fallen into

---

* There doesn't seem to be a good word for 'to be altruistic'. To altru?

a lake, people do not ask 'Excuse me, sir, but how closely related are you to me? Are you, by any chance, a close relative?' before diving in to rescue them. If they are the sort of people who dive in, they do so whoever has fallen into the lake. If not, they don't. Mostly. A clear exception arises when a child falls in; even if they can't swim its parent is then very likely indeed to plunge in to the rescue, but probably would not do so for someone else's child, and even less so for an adult. So the genetic orthodoxy does have a certain amount going for it.

Not much, though. Fisher's mathematics is rather old-fashioned, and it rests on a big – and very shaky – modelling simplification.* It represents a species by its gene-pool, where all that matters is the proportion of organisms that possess a given gene. Instead of comparing different strategies that might be adopted by an organism, it works out what strategy is best 'on average'. And inasmuch as individual organisms are represented within its framework at all, which they are only as contributors to the gene-pool, it views competition between organisms as a direct 'me versus thee' choice. A bird that eats seeds is up against a bird that eats worms in a head-to-head struggle for survival, like two tennis-players … and may the best bird win.

This is a bean-counting analysis performed with a bean-counting mentality. The bird with the most beans (energy from seeds or worms, say) survives; the other does not.

From a complex system viewpoint, evolution isn't like that at all. Organisms may sometimes compete directly – two birds tugging at the same worm, for instance. Or two baby birds in the nest, where direct competition can be fierce and fatal. But mostly the competition is indirect – so indirect that 'compete' just isn't the right word. Each individual bird either survives, or not, against the background of everything else, including the other birds. Birds A and B do not go head-to-head. They compete against each other only in the sense that *we* choose to compare how A does with how B does, and declare one of them to be more successful.

It's like two teenagers taking driving tests. Maybe one of them is in

* In Fisher's day, this simplification was a great idea, because it made it possible to do the sums. Nowadays, it's a bad idea, for the same reason. You can *do* them, but you can't put any faith in the answers.

the UK and the other is in the USA. If one passes and the other fails, then we can declare the one who has passed to be the 'winner'. But the two teenagers don't even know they are competing, for the very good reason that they're not. The success or failure of one has no effect on the success or failure of the other. Nevertheless, one gets to drive a car, and the other doesn't.

The driving-test system works that way, and it doesn't matter that the American test is easier to pass than the British one (as we can attest from personal experience). Evolutionary 'competition' mostly works like the driving test, but with the added complication that just occasionally it really is more like a tennis match.

From this point of view, evolution is a complex system, with organisms as entities. Which organisms survive to reproduce, and which do not, are system-level properties. They depend as much on context (American driving test versus British) as on the internal features of the individuals. The survival of a species is an emergent feature of the whole system, and no simple short-cut computation can predict it. In particular, computations based on the frequencies of genes in the gene-pool can't predict it, and the alleged explanation of altruism by gene-frequencies is unconvincing.

Why, then, does altruism arise? An intriguing answer was given by Randolph Nesse in the magazine *Science and Spirit* in 1999. In a word, his answer is 'overcommitment'. And it is a refreshing and much-needed alternative to bean-counting.

We have said more than once that humans are time-binders. We run our lives not just on what is happening now, but on what we think will happen in the future. This makes it possible for us to commit ourselves to a future action. 'If you fall sick, I will look after you.' 'If an enemy attacks you, I will come to your aid.' Commitment strategies change the face of 'competition' completely. An example is the strategy of 'mutual assured destruction' as a deterrent for nuclear war: 'If you attack me with nuclear weapons, I will use mine to destroy your country completely.' Even if one country has many more nuclear weapons, which on a bean-counting basis means that it will 'win', the commitment strategy means that it can't.

If two people, tribes or nations make a pact, and agree to commit support to each other, then they are both strengthened, and their survival prospects increase. (Provided it's a sensible pact. We leave you to invent scenarios where what we've just said is wrong.) Ah, yes, that's all very well, but can you trust the other to keep to the agreement? We have evolved some quite effective methods for deciding whether or not to trust someone. At the simplest level, we watch what they do and compare it to what they say. We can also try to find out how they have behaved in similar circumstances before. As long as we can get such decisions right most of the time, they offer a substantial survival advantage. They improve how well *we* do, against the background of everything else. Comparison with others is irrelevant.

From a bean-counter's point of view, the 'correct' strategy in such circumstances is to count how many beans you gain by committing yourself, compare that to how many you gain by cheating, and see which pile of beans is biggest. From Nesse's point of view, that approach doesn't amount to a hill of beans. The whole calculation can be sidestepped, at a stroke, by the strategy of overcommitment. 'Stuff the beans: I guarantee that I will commit myself to you, *no matter what*. And you can trust me, because I will prove to you, and keep proving it every day that we live, that I am committed at that level.' Overcommitment beats the bean-counters hands down. While they're trying to compare 142 beans with 143, overcommitment has wiped the floor with them.

Nesse suggests that such strategies have had a decisive effect in shaping our extelligence (though he doesn't use that word):

> Commitment strategies give rise to complexities that may be a selective force that has shaped human intelligence. This is why human psychology and relationships are so hard to fathom. Perhaps a better understanding of the deep roots of commitment will illuminate the relationships between reason and emotion, and biology and belief.

Or, to put it another way: perhaps that's what gave us an edge over the Neanderthals. Though it would be difficult to find a scientific test for such a suggestion.

When humans overcommit in this manner, we call it 'love'. There is far more to love than the simple scenario just outlined, of course, but one feature is common to both: love counts not the cost. It doesn't *care* about who gets the most beans.* And by refusing to play the bean-counters' game, it wins outright. Which is a very religious, spiritual and uplifting message. *And* sound evolutionary sense. What more could we ask?

Quite a bit, actually, because now it all starts to get nasty. The reasons, however, are admirable. Every culture needs its own Make-a-Human kit, to build into the next generation the kind of mind that will keep the culture going – and, recursively, ensure that the next generation does the same for the one that comes after that. Rituals fit very readily into such a kit, because it is easy to distinguish Us from Them by the rituals that We follow but They don't.† It is also an excellent test of a child's willingness to obey cultural norms by insisting that they carry out some perfectly ordinary task in an unnecessarily prescribed and elaborate manner.

Now, however, the priesthood has got its ideological toe in the cultural doorway. Rituals need someone to organise them, and to elaborate them. Every bureaucracy builds itself an empire by creating unnecessary tasks and then finding people to carry them out. A

---

* Altruism, cooperation and love among humans are not the only examples of evolutionary overcommitment … as the Librarian well knows. A banana is *much* better suited to being eaten by an orangutan than it needs to be. The rest of the fruit kingdom doesn't come close. What's in it for other fruit, like the tomato, is that its seeds pass through the animal and are dispersed, complete with a built-in packet of fertiliser. A bean-counting tomato could reduce its level of suitability and still ensure that its seeds, rather than those of the competition, were propagated (the juiciest tomatoes used to be from the plants growing at the sewage farm …). But an over-committed banana avoids the need to test such fine points. By going over the top, losing its seed-producing capacity entirely and relying on humans to propagate it, it ensures that it wins so comfortably that no competitor even gets a look in.

† … which can be applied so overpoweringly that the people who aren't Us aren't anything. See the Imperial China parody – the Agatean Empire – in *Interesting Times*, and a number of Roundworld cultures, too. Being Them is quite a step up by comparison.

crucial task here is to ensure that members of the tribe or village or nation really do obey the norms and carry out the rituals. There has to be some sanction to make sure that they do, even if they're free-thinking types who'd rather not. Because everything is founded on an ontically dumped concept, reference to reality has to be replaced by belief. The less testable a human belief is, the more strongly we tend to hold on to it. Deep down we recognise that although not being testable means that disbelievers can't prove we're wrong, it also means that we can't prove we're right. Since we *know* that we are, that sets up a tremendous tension.

Now the atrocities begin. Religion slides over the edge of sanity, and the result is horrors like the Spanish Inquisition. Think about it for a moment. The priesthood of a religion whose central tenet was universal love and brotherhood systematically inflicted appalling tortures, sick and disgusting things, on innocent people who merely happened to disagree about *minor* items of belief. This is a massive contradiction and it demands explanation. Were the Inquisitors evil people who knowingly did evil things?

Small Gods, one of the most profound and philosophical of the Discworld novels, examines the role of belief in religions, and Discworld undergoes its own version of the Spanish Inquisition. One twist is that on Discworld, there is no lack of gods; however, few of them have any great significance:

> There are billions of gods in the world. They swarm as thick as herring roe. Most of them are too small to see and never get worshipped, at least by anything bigger than bacteria, who never say their prayers and don't demand much in the way of miracles.
>
> They are the small gods, the spirits of places where two ant trails cross, the gods of microclimates down between the grass roots. And most of them stay that way.
>
> Because what they lack is *belief.*

Small Gods is the story of one rather larger god, the Great God Om, who manifests himself to a novice monk called Brutha, in the Citadel at the heart of the city of Kom in the lands between the deserts of

Klatch and the jungles of Howondaland.

Brutha's attitude to religion is a very personal one. He runs his own life by it. In contrast, Deacon Vorbis believes that the role of religion is to run everybody else's life. Vorbis is head of the Quisition, whose role is 'to do all those things that needed to be done and which other people would rather not do'. Nobody ever interrupts Vorbis to ask what he is thinking about, because they are scared stiff that the answer will be 'You'.

The Great God's manifestation takes the form of a small tortoise. Brutha finds this hard to believe:

> I've seen the Great God Om … and he isn't tortoise-shaped. He comes as an eagle, or a lion, or a mighty bull. There's a statue in the Great Temple. It's seven cubits high. It's got bronze on it and everything. It's trampling infidels. You can't trample infidels when you're a tortoise.

Om's power has waned because of a lack of belief. He tests his strength by silently cursing a beetle, but it makes no difference and the insect plods away unperturbed. He curses a melon unto the eighth generation, but with no evident effect. He inflicts a plague of boils on it, but all it does is sit there, slowly ripening. He vows that when he returns to his rightful state, the Tribes of Beetle and Melons will regret not responding. For on Discworld, the size of a god is determined by the strength, and amount, of belief in him (or her, or it). Om's church had become so corrupt and powerful that the fearful belief of the common people had been transferred to the church itself – it's very *easy* to believe in a red-hot poker – and only Brutha, simple soul, still truly believes. No god ever dies, because there is always some tiny pocket of belief remaining somewhere in the world, but a tortoise is pretty much as low as you can get.

Brutha is going to become the Eighth Prophet of Om. (His grandmother would have made it two generations before, but she was a woman, and narrative imperative forbids female prophets.) Vorbis's job is to ensure that all Omnians remain true to the teachings of the Great God Om, which is to say, they do what Vorbis tells them. The presence on the Discworld of the god itself, causing changes to all the old teachings and generally making trouble, is not greatly to Vorbis's taste. Neither is the presence of a genuine prophet. Vorbis is faced with the inquisitor's spiritual dilemma, and resolves it in the time-honoured manner of the Spanish Inquisition (which, basically, is to tell oneself that torturing people is fine because it's for their own good, in the long run).

Brutha has a much simpler vision of Omnianism: it is something for individuals to live by. Vorbis shows Brutha a new instrument that he has had made: an iron turtle upon which a man or woman can be spreadeagled, with a firebox inside. The time it takes for the iron to heat up will give them plenty of time to reflect on their heresies. In a flash of prophecy, Brutha realises that its first victim will be himself. And in due course, he finds himself chained to it, and uncomfortably warm, with Vorbis watching over him, gloating. Then the Great God Om intervenes, dropped from the talons of an eagle.

> One or two people, who had been watching Vorbis closely, said later
> that there was just time for his expression to change before two
> pounds of tortoise, travelling at three metres per second, hit him
> between the eyes.
>
> It was a revelation.
>
> And that does something to people watching. For a start, they
> believe with all their heart.

The Great God Om now is truly great. He rises over the Temple, a billowing cloud shaped like eagle-headed men, bulls, golden horns, all tangled and fused into one another. Four bolts of fire whir out of the cloud and burst the chains that fastened Brutha to the iron turtle. The Great God declares Brutha to be Prophet of Prophets.

The Great God gives Brutha the opportunity to make some Commandments. The Prophet declines, having decided that 'You should do things because they're right. Not because gods say so. They might say something different another time'. And he tells Om that there will be no Commandments unless the god agrees to obey them, too.

Which is a new thought, for a god.

Small Gods has many wise words to say about religion and belief, and it makes the point that in their own terms the Inquisitors believe they are doing good. Fyodor Dostoyevsky's *The Brothers Karamazov* has a scene in which the Grand Inquisitor encounters Christ, and explains his point of view, including why Christ's renewed message of universal love couldn't have come at a worse time and will only cause trouble. Just as the presence of Brutha, a genuine prophet, was not at all to the liking of Deacon Vorbis.

The Spanish Inquisitors' justification of their actions was philosophically convoluted. The purpose of their tortures was straightforward: it was to save a sinner from eternal damnation. The tortures of Hell would be far worse than anything that the Inquisitors could inflict in this world, and they would never stop. So *of course* they were justified in using any means whatsoever to save the poor soul from destruction. They therefore believed that their actions were justified, and in accordance with Christian principles. Not to act would have been to leave the person concerned in danger of the terrible fires of Hell.

Yes, but what if they were wrong in this belief? This is the convoluted bit. They weren't quite sure about their religious position. What were the rules? If they failed to convert one tortured heretic, would the *Inquisitors* burn forever? If they converted one heretic, would their souls be guaranteed a place in Heaven? The Inquisitors believed that by inflicting pain and terror without knowing the rules, they risked their own mortal souls. If they were wrong, it was they who would be immersed in the eternal flames. But they were willing to risk this enormous spiritual danger, to take upon themselves all of the consequences of their actions, should they turn out to be wrong. See how incredibly magnanimous they were being, even as they burned people alive and hacked them limb from limb with red-hot knives …

Clearly something is wrong. Dostoyevsky solves his own narrative problem by having Christ respond the way his own teachings would lead him to: he kisses the Inquisitor. This is an answer, of a kind, but it doesn't satisfy our analytical instincts. There is a logical flaw in the Inquisitors' position: what is it?

It's very simple. They have thought about what happens if their belief that their actions are justified is wrong – but only within the frame of their religion. They have not asked themselves what their position would be if their religious beliefs are false, if there is no Hell, no eternal damnation, no fire and brimstone. Then their justification would fall to bits.

Of course, if their religion is wrong, then its doctrine of brotherly love could also be wrong. It doesn't have to be: some parts might be fine, others nonsense. But to the Inquisitors it is all of one piece, it stands or falls as a whole. If they are wrong about their religion, then there is no sin, no God, and they can cheerfully torture people if they want to. It really is a nasty philosophical trap.

This is the kind of thing that happens when a big, powerful priesthood latches on to what started as one person's awe at the universe. It is what happens when people construct elaborate verbal traps for themselves, trip over the logic, and fall headlong into them. It is where Holy Wars come from, where neighbour can inflict atrocity on neighbour merely because this otherwise reasonable person goes to a church with a round tower instead of a square one. It is the attitude that Jonathan Swift caricatured in *Gulliver's Travels*, with the conflict between the big-endians and the little-endians, over which end of an egg to slice into when eating it. It is, perhaps, why so many people today are turning to unorthodox cults in an effort to find a home for their own spirituality. But cults run the same risk as the Inquisition. The only safe home for one's personal spirituality is oneself.

TWENTY-ONE

THE NEW SCIENTIST

 THERE WAS SOMETHING CALLED, as far as Ponder could work out, psyence. All his expertise as a reader of invisible writings was needed to get a grip on this idea – L-space was very hazy about the future of this world.

'As far as I can tell,' he reported, 'it's a way of making up stories that work. It's a way of finding things out and thinking about them ... *psy*-ence, you see? "Psy" means "mind" and "ence" means, er, *esness*. It works on Roundworld in the way magic does at home.'

'Useful stuff, then,' said Ridcully. 'Anyone doing it?'

'Hex is going to try to take us to what appear to be practical examples of it,' said Ponder.

'Time travel *again?*' said the Dean.

The white circle appeared on the floor ...

... and on the sand, and vanished.

The wizards looked around.

'All right, then,' said Ponder. 'So ... dry climate, evidence of agriculture, fields of crops, irrigation ditches, naked man turning a handle, man staring at us, man screaming and running away ...'

Rincewind stepped down into the ditch and inspected the pipe-like device the man had been turning.

'It's just a water-lifting screw,' he announced. 'I've seen a lot of them. You turn the handle, water is screwed out of the ditch, goes up the thread inside and spills out of the top. The screw makes a sort of line of travelling buckets inside the tube. There's nothing *special* about it.

It's just basic ... stuff.'

'Not psyence, then?' said Ridcully.

'You tell me, sir,' said Rincewind.

'Psyence is quite a difficult concept,' said Ponder. 'But I think perhaps tinkering with this thing to make it more efficient might be psyence?'

'Sounds like engineering,' said the Lecturer in Recent Runes. 'That's where you try and make it in different ways to see if any of them are better.'

'The Librarian did turn up one book, very grudgingly,' said Ponder, pulling it out of his pocket.

It was called *Basic Science for Schools, pub. 1920*.

'They've spelt it wrong,' said Ridcully.

'And it's not very helpful,' said Ponder. 'There's quite a lot of what looks like alchemy. You know, mixing stuff up to see what happens.'

'Is that all it is, then?' said the Archchancellor, leafing through the book. 'Hold on, hold on. *Alchemy* is, at bottom, all about the alchemist. His books tell him all the stuff he's got to do in order make things work – what to wear, when to wear it, that sort of thing. It's very personal.'

'And?' said the Lecturer in Recent Runes.

'Hark at this,' said Ridcully. 'There's no invocations, nothing to tell you what to wear or what phase of the moon it should be. Nothing important. It just says here "A clean beaker was taken. To this was added 20 grammes" – whatever they are – "of copper sulphate" ...' He stopped.

'Well?' said the Lecturer in Recent Runes.

'Well, who did the taking? Who added the stuff? What's going on *here?*'

'Perhaps it's trying to say that it doesn't matter who does it?' said Ponder. He'd already glanced at the book, and felt that the perfectly ordinary ignorance he'd had just before opening it had been multiplied several times by page ten.

'Anyone can do it?' shouted Ridcully. 'Science is incredibly important but *anyone* can do it? And what's this?'

He held the book open for all to see, his finger pointing at an illustration. It showed a drawing of an eye, side on, to one side of the apparatus.

'Perhaps it's a God of Science?' Rincewind suggested. 'Watching to see who keeps taking things?'

'So ... science is done by anyone,' said Ridcully, 'and most of the equipment is stolen and it's all watched by a giant eyeball?'

As one wizard, they looked around, guiltily.

'There's just us,' said Ponder.

'Then this isn't science,' said Ridcully. 'No giant eyeball visible. Anyway, we can *see* it isn't science. It's just engineering. Any bright lad could have built it. It's obvious how it works.'

'How *does* it work?' said Rincewind.

'Very simply,' said Ridcully. 'The screw goes round and round and the water comes out here.'

'Hex?' said Ponder, and held out his hand. A large volume appeared in it. It was slim, full of colourful pictures, and entitled *Great Moments in Science*. It hadn't escaped his notice that when Hex or the Librarian wanted to explain something to the wizards they used a children's book.

He flicked through the passages. Big pictures, big writing.

'Ah,' he said. 'Archimedes invented this. He was a philosopher. He's also famous because one day, when he got into his bath, it overflowed. It says here this gave him an idea—'

'Buy a bigger bath?' said the Dean.

'Philosophers are always having ideas in the bath,' said Ridcully. 'All right, if we've got nothing else to go on ...'

'Gentlemen, please?' pleaded Ponder. 'Hex, take us to Archimedes. Oh, and give me a towel ...'

'Nice place,' said the Dean, as the wizards sat on the sea wall, staring out at the wine-dark sea. 'I can feel the sea air doing me good. Anyone for more wine?'

It had been quite an interesting day. But, Ponder asked, had it been science? There was a pile of books beside him. Hex had been busy.

'Must have been science,' said Ridcully. 'King gave your man a problem. How to tell if the crown was all gold. He was thinking about it.

Water sloshed out of bath. He leaped out, we handed him a towel, and then he worked out that ... what was it?'

'The apparent loss of weight of a body totally or partially immersed in a liquid is equal to the weight of the liquid it displaces,' said Ponder.

'Right. And he sees it doesn't just work with bodies, it works with crowns, too. A few tests, and bingo, science,' said Ridcully. 'Science is just working things out. And paying attention. And hoping there's someone around to dry you off.'

'I'm not ... *exactly* sure that's all there is to it,' said Ponder. 'I've been doing some reading and even people who do science don't seem clear about what it is. Look at Archimedes, for example. Is a bright idea enough? Is it science if you just solve problems? Is that science, or what you get before you have science?'

'Your book of Great Moments calls him a scienter,' Ridcully pointed.

'Scientist,' Ponder corrected him. 'But I'm not sure about that, either. I mean, that sort of thing happens a lot. People always like to believe that what they're doing has been hallowed by history. Supposing men found out how to fly. They'd probably say "Early experimenters with man-powered flight included Gudrun the Idiot, who leaped off the clock tower in Pseudopolis after soaking his trousers in dew and gluing swan feathers to his shirt" when in fact he wasn't an early aviator—'

'—he was a late idiot?' said Rincewind.

'Exactly. It's like with wizards, Archchancellor. You can't just call yourself a wizard. Other wizards have to agree that you're a wizard.'

'So you can't have just one scientist, but you can have two?'

'It appears so, Archchancellor.'

Ridcully lit his pipe. 'Well, mildly entertaining though it is to watch philosophers having a bath, can we simply ask Hex to find us a scientist who is definitely a scientist and who is regarded by other scientists as a scientist? Then all we have to do is find out if what he's doing is any use to us. We don't want to be all day at this, Stibbons.'

'Yes, sir. Hex, we—'

They were in a cellar. It was quite large, which was just as well because several of the wizards fell over upon landing. When they had picked themselves up and all found the right hat, they saw ...

... something familiar.

'Mr Stibbons?' said Ridcully.

'I don't understand ...' muttered Ponder. But it really *was* an alchemical laboratory. It smelled like one. Moreover, it looked like one. There were the big heavy retorts, the crucibles, the fire ...

'We *know* what alchemists are, Mr Stibbons.'

'Yes, er, I'm sorry, sir, something seems to have gone wrong ...' Ponder held out his hand. 'Book, please, Hex.'

A small volume appeared.

'"Great Men Of Science No.2",' Ponder read. 'Er ... if I can just take a quick look inside, Archchancellor ...'

'I don't think that will be necessary,' said the Dean, who had picked up a manuscript that was on the table. 'Listen to this, gentlemen: " ... The spirit of this earth is yᵉ fire in wᶜʰ Pontanus digests his feculent matter, the blood of infants in wᶜʰ yᵉ ☉ & ☽ bath themselves, the unclean green Lion wᶜʰ, saith Ripley, is yᵉ means of joyning yᵉ tinctures of ☉ and ☽, the broth wᶜʰ Medea poured on yᵉ two serpents, the Venus by meditation of wᶜʰ ☉ vulgar and the ☿ of 7 eagles saith Philalethes must be decocted ..." yada yada yada.'

He thumped the manuscript on to the table.

'Genuine alchemical gibberish,' he said, 'and I don't like the sound of it. What's "feculent" mean? Do we dare find out? I think *not*.'

'Er ... the man who apparently lived here is described as a giant amongst scientists ...' muttered Ponder, leafing though the booklet.

'Really?' said Ridcully, with a dismissive sniff. 'Hex, *please* take us to a *scientist*. We don't mind where he is. Not some dabbler. We want someone who embodies the very essence of *science*.'

Ponder sighed, and dropped the booklet on to the ground.

The wizards vanished.

For a moment the book lay on the floorboards, front cover upwards showing its title: *Great Men of Science No. 2: Sir Isaac Newton*. Then it, too, vanished.

There was a thunderstorm grumbling in the distance, and black clouds hung over the sea. The wizards were back on a beach again.

'Why is it always beaches?' said Rincewind.

'Edges,' said Ridcully. 'Things happen on the edges.'

They had been happening here. At first glance the place looked like a shipyard that had launched its last ship. Large wooden constructions, most of them in disrepair, littered the sand. There were a few shacks, too, also with that hopeless look of things abandoned. There was nothing but desolation.

And an oppressive, silence. A few sea birds cried and flew away, but that only left the world to the sound of waves and the footfalls of the wizards as they approached the shacks.

At which point, another sound became apparent. It was a rhythmical cracking, a *khss ... khss ... khss* behind which it was just possible to hear voices raised in song; the singers sounded as if they were far away and at the bottom of a tin bath.

Ridcully stopped outside the largest shack, from which the sound appeared to be issuing.

'Rincewind?' he said, beckoning. 'One for you, I think.'

'Yes, yes, all right,' said Rincewind, and entered with extreme caution.

It was dark inside, but he could see workbenches and a few tools, with a forgotten look about them. The shack must have been thrown up quickly. There wasn't even a floor; it had been built directly on the sand.

The singing was coming from a large horn attached to a device on a bench. Rincewind wasn't very good at technical things, but there was a large wheel projecting over the edge of the bench and it was turning slowly, probably because of the small weight, attached to it by string, which was gently descending towards the sand.

'Is everything okay?' said Ridcully, from outside.

'I've found a kind of voice mill,' said Rincewind.

'That's amazing,' said a voice from the shadows. 'That's exactly what my master called it.'

His name, he said, was Niklias the Cretan, and he was very old. And very pleased to see the wizards.

'I come up here sometimes,' he said. 'I listen to the voice mill and remember the old days. No one else comes here. They say it's the abode of madness. And they are right.'

The wizards were sitting around a fire of driftwood, that burned blue with the salt. They were tending to huddle, although they'd never admit it. They wouldn't have been wizards if they couldn't sense the strangeness in the place. It had the same depressing effect on the senses as an old battlefield. It had ghosts.

'Tell us,' said Ridcully.

'My master was Phocian the Touched,' said Niklias, and he said it the way of a man telling a story he'd told many times before. 'He was a pupil of the great philosopher Antigonus, who one day declared that a trotting horse must at all times have at least one foot on the ground, lest it fall over.

'There was much debate about this and my master, being very rich and also being a keen pupil, decided to prove that the philosopher was correct. Oh, dreadful day! For it was then the troubles began …'

The old slave pointed to some derelict woodwork at the far end of the beach.

'That was our test track,' he said. 'The first of four. I helped him build it with my own hands. There was a lot of interest at that time, and many people came to watch the tests. We had hundreds, *hundreds* of slaves lying in rows, peering through little slits at just one tiny area of the track each. It didn't work. They argued about what they had seen.'

Niklias sighed. 'Time, said my master, was important. So I told him about work gangs, and how songs helped us keep time. He was very excited about that, and after some thought we built the voice mill which you have heard. Do not be afraid. There is no magic in it. Sound makes things shake, does it not? Sound in the big parchment horn, which I stiffened with shellac, writes the pattern of the sounds it hears on a warm wax cylinder. We used the weighted wheel to spin the cylinder, and it worked quite well after we devised the rocking-trap mechanism. After that, we used it to inscribe the perfect song, and every dawn before we began work we would sing it with the machine. Hundreds of slaves, all singing in perfect time on this beach. The effect was amazing.'

'I bet it was,' said Ridcully.

'But still it did not work, no matter what we devised. A trotting horse travels too fast. My master told me that we must be able to count in tiny parts of time, and after much thinking we built the toc-toc machine. Would you care to see it?'

It was like the voice mill, but had a much bigger wheel. And a pendulum. And a big pointer. As the big wheel turned very slowly, smaller wheels inside the mechanism spun in a blur, and caused a long pointer to revolve against a white-painted wooden wall, along an arc covered in tiny markers. The whole device was mounted on wheels, and had probably taken four men to move.

'I come and grease it occasionally,' said Niklias, patting the wheel. 'For old time's sake.'

The wizards looked at one another with a tame surmise, which is a wild surmise that had been thought about for a while.

'It's a clock,' said the Dean.

'Pardon?' said Niklias.

'We have something like them,' said Ponder. 'We use them for telling the time.'

The slave looked puzzled. 'For telling the time what?' he said.

'He means, so that we know what time it is,' said Ridcully.

'What ... time ... it ... is ...' muttered the slave, as if trying a square thought in a round mind.

'What hour of the day it is,' said Rincewind, who had run into minds like this before.

'But we can see the sun,' said the slave. 'The toc-toc mechanism does not know where the sun is.'

'Oh, I know ... supposing a baker needed to know how long he should bake his loaves,' said Rincewind. 'Well, with a clock he—'

'How could he be a baker if he did not know how long it takes to bake a loaf?' said Niklias, smiling nervously. 'No, this is a special thing, sirs. It is not for uncursed men.'

'But, but ... you've got a device for recording sound, too!' Ponder burst out. 'You could record the speeches of great thinkers! Why, even after they were dead you could still hear—'

'*Listen* to the voice of people who aren't there?' said Niklias. His face clouded. 'Listen to the *voices* of *dead men?*'

There was silence.

'Do tell us more about the fascinating project to find out if a trotting horse is ever entirely airborne,' said Rincewind, loudly and brightly.

The sun drifted down the sky or, rather, the horizon gradually rose. The wizards hated to think about that. You could lose your balance if you thought about it too much.

'... finally my master came up with a new idea,' said Niklias.

'Another one?' said the Dean. 'Was it better than his idea about dropping a horse from a sling to see if it fell over?'

'Dean!' snapped Ridcully.

'Yes, it was,' said the old slave, who didn't seem to notice the sarcasm. 'We still used the sling, but this time we put it in a very large cart. The bottom of the cart was open, so that the horse's hooves just touched the ground. Are you following me? And then – and this is the clever part, I felt – my master arranged that the cart was *pulled by four trotting horses.*'

He sat back, giving them a pleased look, as if expecting praise.

The Dean's expression slowly changed.

'Eureka!' he said.

'I've got a towel in my—' Rincewind began.

'No, don't you see? If the cart is being pulled forward then whatever the horse does, the ground is disappearing *backwards*. So if you've got a trained horse and you can get it to trot while it's in the harness ... you designed the cart so that the pulling horses were offset, so that the supported horse was trotting over unmarked sand?'

'Yes!' beamed Niklias.

'And you raked the sand so that the prints showed up?'

'Yes!'

'Then whenever the horse touched the ground and the hoof was stationary relative to the ground, the ground would in fact be moving, and you'd get a smeared print, and if you carefully measured the total length of the ground covered during the trot, and added up the total of all the smears, and found that they were less than the total length of the track, then—'

'You'd be doing it wrong,' said Ponder.

'Yes!' said Niklias, delightedly. 'That's what we found!'

'No, of course it's right,' said the Dean. 'Listen: when the hoof is stationary—'

'It's moving backwards relative to the horse at the same speed that the horse is moving forward,' said Ponder. 'Sorry.'

'No, listen,' the Dean protested. 'It *must* work, because when the ground *isn't* moving—'

Rincewind groaned. Any minute now all the wizards would express an opinion, and none of them would listen to anyone else. And here it came ...

'Are you telling us parts of the horse are actually going *backwards?*'

'Perhaps if we pulled the cart in the opposite direction—'

'The hoof would definitely be stationary, look, because if the ground was moving forward—'

'It's no different than it would be if the horse was trotting all by itself! Look, supposing the cart and all the other horses were invisible—'

'You're all wrong, you're all wrong! If the horse was ... no, wait a moment—'

Rincewind nodded to himself. The wizards were entering the special fugue state known as Hubbub, where no-one was going to be allowed to finish a sentence because someone else would drown them out. It was how the wizards decided things. In all likelihood, in this case it would result in them deciding that the horse should, logically, end up at one end of the beach, while all its feet were up at the other end.

'My master Phocian said we should try it, and the hooves just left hoofprints,' said Niklias the Cretan, when the argument had died away through lack of breath. 'Then we tried moving the beach under the horse ...'

'How?' said Ponder.

'We built a long flat barge, filled it full of sand and tried it in the lagoon,' said the slave. 'We suspended the horse from a gantry. Phocian felt we were getting somewhere when we moved the barge forward at twice the speed of the horse, but the beast kept trying to keep up ... and then there was the night of the big storm and the barge was sunk. Oh, those were a few busy months. We lost four horses and Nosios the Carpenter was kicked in the head.' The smile faded. 'And then ... and then ...'

'Yes?'

'... something terrible happened.'

The wizards leaned forward.

'... Phocian designed the fourth test. It's over there. Not much to see now, of course. People stole all the heavy cloth of the Endless Road and a lot of the woodwork, too.' The slave sighed. 'It was Hades to build and took many months to get right but, in short, it worked like this. We used a huge roll of heavy white cloth, which we rolled off one huge spindle and on to the other. Believe me, sirs, even that took some doing, and the work of forty slaves. At the place where the horse was to be suspended, we stretched the cloth tight over a shallow trough of powdered charcoal, so that a little weight on the cloth would press it down on to the stuff ...'

'Aha,' said the Dean. 'I think I can see this one ...'

Niklias nodded. 'My master commanded many changes before the device functioned to his satisfaction ... many gears and rollers and cranks, much rebuilding of strange mechanisms, much profanity which, I have no doubt, the gods noted. But finally we suspended the well-trained horse in its sling and the rider urged it into a trot as the cloth rolled beneath. And, yes, afterwards, oh sad that day, we measured the length of the cloth where the horse had trotted and the length of the smears of charcoal where a hoof had pressed on the cloth and ... I hardly dare say it, even now, the total length of the second was to the length of the first was as four is to five.'

'So for a fifth of the time all hooves were in the air!' said the Dean. 'Well done! I love a puzzle!'

'No, it was not well done!' shouted the slave. 'My master ranted! We did it again and again! And it was always the same!'

'I don't quite see the problem—' Ridcully began.

'He tore at his hair and raved at us, and most of the men fled! And then he went and sat in the waves on the shore, and after a long while

I dared to go and speak to him, and he turned hollow eyes on me and said, "Great Antigonus is wrong. I proved him wrong! Not by thoughtful dispute, but by gross mechanical contrivances! I am ashamed! He is the greatest of philosophers! He had told us that the sun goes around the world, he had told us how the planets move! And if he is wrong, what is right? What have I done? I have squandered the wealth of my family. What fame is there for me now? What cursed work shall I do next? Should I steal the colours from a flower? Shall I say to everyone, 'What you think is right, is not right'? Shall I weigh the stars? Shall I plumb the utter depths of the sea? Shall I ask the poet to measure the width of love and the direction of pleasure? What have I made of myself … " and he wept.'

There was silence. None of the wizards moved.

Niklias settled down a little. 'And then he bade me go back and he told me to take the little money that was left. In the morning he was gone. Some say he fled to Egypt, some say to Italy. But for myself, I think he did indeed plumb, at the last, the depth of the sea. For I do not know what he was, or what he had become. And presently people came and tore down most of the engines.'

He shifted his weight and looked at the remains of the strange devices, skeletal against the livid sunset. There was something wistful in his expression.

'No one comes now,' he said. 'Hardly anyone at all. This is where the Fates struck and the gods laughed at men. But I remember how he wept. And so I remain, to tell the story.'

TWENTY-TWO
THE NEW
NARRATIVIUM

THE WIZARDS HAVE BEEN TRYING to find some 'psyence' in Roundworld, but it is proving even more elusive than the correct spelling.

They are having problems because they are tackling a difficult question. There isn't a simple definition of 'science' that really captures what it is. And it's not the sort of thing that comes into existence at a single place and time. The development of science was a process in which non-science slowly *became* science. The two ends of the process are easily distinguished, but there's no special place in between where science suddenly came into being.

These difficulties are more common than you might expect. It is almost impossible to define a concept precisely – think of 'chair', for example. Is a large beanbag a chair? It is if the designer says it's a chair and someone uses it to sit on; it's not if a bunch of kids are throwing it at each other. The meaning of 'chair' does not just depend on the thing to which it is being applied: it also depends on the associated context. And as for processes in which something gradually changes into something else … well, we're never comfortable with those. At what stage in its life does a developing embryo become a human being, for instance? Where do you draw the line?

You don't. If the end of a process is qualitatively different from the start, then something changes in between. But it need not be at a specific *place* in between, and if the change is gradual, there *isn't* a line. Nobody thinks that when an artist is painting something, there is one special stroke of the brush at which it turns into a picture. And nobody asks 'Whereabouts in that particular brushstroke does the change take place?' At first there is a blank canvas, later there's a picture, but there isn't a well-defined moment at which one ceases and the other begins. Instead, there is a long period of neither.

We accept this about a painting, but when it comes to more emotive processes like embryos becoming human beings, a lot of us still feel the need to draw a line. And the law encourages us to think like that, in black and white, with no intervening shades of grey. But that's not how the universe works. And it certainly didn't work like that for science.

To complicate things even further, important words have changed their meaning. An old text from 1340 states that 'God of sciens is lord', but there the word* 'sciens' means 'knowledge', and the phrase is saying that God is lord of knowledge. For a long time science was known as 'natural philosophy', but by 1725 the word 'science' is being used in essentially its modern form. The word 'scientist', however, seems to have been invented by William Whewell in his 1840 *The Philosophy of the Inductive Sciences* to describe a practitioner of science. But there were scientists before Whewell invented a word for them, otherwise he wouldn't have needed a word, and there was no science when God was lord of knowledge. So we can't just go by the words people use, as if words never change their meanings, or as if things can't exist before we have a word for them.

But surely science goes back a long, long way? Archimedes was a scientist, wasn't he? Well, it depends. It certainly looks to us, now, as if Archimedes was doing science; indeed we have reached back into history, picked out some of his work (especially his buoyancy principle) and called it science. But he wasn't doing science *then*, because the context wasn't suitable, and his mind-set was not 'scientific'. We see him with hindsight; we turn him into something we recognise, but he wouldn't.

Archimedes made a brilliant discovery, but he didn't test his ideas like a scientist would now, and he didn't investigate the problem in a

---

* Other recorded spellings are cience, ciens, scians, scyence, sience, syence, syens, syense, scyense. Oh, and science. Naturally, the wizards have invented another one.

genuinely scientific way. His work was an important step along the path to science, but one step is not a path. And one thought is not a way of thinking.

What about the Archimedean screw? Was that science? This wonderful device is a helix that fits tightly inside a cylinder. You place the cylinder at a slant, with the bottom end in water; turn the helix, and after a while water comes out at the top. It is generally believed that the famous Hanging Gardens of Babylon were watered using massive Archimedean screws. How it works is more subtle than Ridcully imagines: in particular, the screw ceases to work if it is held at too steep an angle. Rincewind is right: an Archimedean screw is like a series of travelling buckets, separate compartments with water in them. Because they are separate, there is no continuous channel for the water to flow away along. As the screw turns, the compartments move up the cylinder, and the water has to go with them. If you hold the cylinder at too steep a slope, all the 'buckets' merge, and the water no longer climbs.

The Archimedean screw surely counts as an example of ancient Greek technology, and it illustrates their possession of engineering. We tend to think of the Greeks as 'pure thinkers', but that's the result of selective reporting. Yes, the Greeks were renowned for their (pure) mathematics, art, sculpture, poetry, drama and philosophy. But their abilities did not stop there. They also had quite a lot of technology. A fine example is the Antikythera mechanism, which is a lump of corroded metal that some fishermen found at the bottom of the Mediterranean Sea in 1900 near the island of Antikythera.* Nobody took much notice until 1972, when Derek de Solla Price had the lump X-rayed. It turned out to be an orrery: a calculating device for the movements of the planets, built from 32 remarkably precise cogwheels. There was even a differential gear. Before this gadget was discovered, we simply didn't know that the Greeks had possessed that kind of technological ability.

---

* So called because it is near the larger island of Kythera. This is 'anti' = near, not 'anti' = opposed to. Though, metaphorically, the two usages are close. Think about the meaning of 'opposed to'. And 'against'.

We still don't understand the context in which the Greeks developed this device; we have no idea where these technologies came from. They were probably passed down from craftsman to craftsman by word of mouth – a common vehicle for technological extelligence, where ideas need to be kept secret *and* passed on to successors. This is how secret craft societies, the best known being the freemasons, arose.

The Antikythera mechanism was Greek engineering, no question. But it wasn't science, for two reasons. One is trivial: technology isn't science. The two are closely associated: technology helps to advance science, and science helps to advance technology. Technology is about making things work without understanding them, while science is about understanding things without making them work.

Science is a *general* method for solving problems. You're only doing science if you know that the method you're using has much wider application. From those written works of Archimedes that still survive, it looks as if his main method for inventing technology was mathematical. He would lay down some general principles, such as the law of the lever, and then he would think a bit like a modern engineer about how to exploit those principles, but his derivation of the principles was based on logic rather than experiment. Genuine science arose only when people began to realise that theory and experiment go hand in hand, and that the combination is an effective way to solve lots of problems and find interesting new ones.

Newton was definitely a scientist, by any reasonable meaning of the word. But not all the time. The mystical passage that we've quoted, complete with alchemical symbols* and obscure terminology, is one that he wrote in the 1690s after more than twenty years of alchemical experimentation. He was then aged about 50. His best work, on mechanics, optics, gravity, and calculus, was done between the ages of 23 and 25, though much of it was not published for decades.

Many elderly scientists go through what is sometimes called a 'philosopause'. They stop doing science and take up not very good

* The symbols have the following meanings: ☉ = Sun, ☽ = Moon, ☿ = Mercury.

philosophy instead. Newton really did investigate alchemy, with some thoroughness. He didn't get anywhere because, frankly, there was nowhere to go. We can't help thinking, though, that if there had been somewhere, he would have found the way.

We often think of Newton as the first of the great rational thinkers, but that's just one aspect of his remarkable mind. He straddled the boundary between old mysticism and new rationality. His writings on alchemy are littered with cabbalistic diagrams, often copied from early, mystical sources. He was, as John Maynard Keynes said in 1942, 'the last of the Magicians ... the last wonder-child to whom the Magi could do sincere and appropriate homage'. What confuses the wizards is an accident of timing – well, we must confess that it is actually a case of narrative imperative. Having homed in on Newton as the epitome of scientific thinking, the wizards happen to catch him in post-philosopausal mode. Hex is having a bad day, or perhaps is trying to tell them something.

If Archimedes wasn't a scientist and Newton was only one some-times, just what is science? Philosophers of science have isolated and defined something called the 'scientific method', which is a formal summary of what the scientific pioneers often did intuitively. Newton followed the scientific method in his early work, but his alchemy was bad science even by the standards of his day, when chemists had already moved on. Archimedes doesn't seem to have followed the sci-entific method, possibly because he was clever enough not to need it.

The textbook scientific method combines two types of activity. One is experiment (or observation – you can't experiment on the Big Bang but you can hope to observe traces that it left). These provide the real-ity-check that is needed to stop human beings believing something because they want it to be true, or because some overriding authority tells them that it's true. However, there is no point in having a reality-check if it's bound to work, so it can't just be the same observations that you started from. Instead, you need some kind of story in your mind.

That story is usually dignified by the word 'hypothesis', but less formally it is the theory that you are trying to test. And you need a way to test it without cheating. The most effective protection against cheating is to say in advance what results you expect to get when you do a new experiment or make a new observation. This is 'prediction', but it may be about something that has already happened but not yet been observed. 'If you look at red giant stars in this new way then you will find that a billion years ago they used to ...' is a prediction in this sense.

The most naïve description of the scientific method is that you start with a theory and test it by experiment. This presents the method as a single-step process, but nothing could be further from the truth. The real scientific method is a recursive interaction between theory and experiment, a complicity in which each modifies the other many times, depending on what the reality-checks indicate along the way.

A scientific investigation probably starts with some chance observation. The scientist thinks about this and asks herself 'why did that happen?' Or it may be a nagging feeling that the conventional wisdom has holes in it. Either way, she then formulates a theory. Then she (or more likely, a specialist colleague) tests that theory by finding some other circumstance in which it might apply, and working out what behaviour it predicts. In other words, the scientist designs an experiment to test the theory.

You might imagine that what she should be trying to do here is to design an experiment that will prove her theory is correct.* However, that's not good science. Good science consists of designing an experiment that will demonstrate that a theory is *wrong* – if it is. So a large part of the scientist's job is not 'establishing truths', it is trying to shoot down the scientist's own ideas. And those of other scientists. This is what we meant when we said that science tries to protect us against believing what we want to be true, or what authority tells us is true. It doesn't always succeed, but that at least is the aim.

---

* On TV news we are repeatedly told about scientists who are 'proving' a theory. Either the people making the programme were trained in media studies and have no idea of how science works, or they were trained in media studies and don't *care* how science works, or they're still wedded to the old-fashioned meaning of the verb 'prove', which means to test. As in the phrase 'the exception proves the rule', which made perfect sense when it was first stated – the exception casts doubt on the rule by 'testing' it *and finding it inadequate* – and makes no sense at all when it is used today to justify ignoring awkward exceptions.

This is the main feature that distinguishes science from ideologies, religions and other belief systems. Religious people often get upset when scientists criticise some aspect of their beliefs. What they fail to appreciate is that scientists are equally critical about their own ideas and those of other scientists. Religions, in contrast, nearly always criticise everything *except* themselves. Buddhism is a notable exception: it emphasises the need to question everything. But that may be going too far to be helpful.

Of course, no real scientist actually follows the textbook scientific method unerringly. Scientists are human beings, and their actions are driven to some extent by their own prejudices. The scientific method is the best one that humanity has yet devised for attempting to overcome those prejudices. That doesn't mean that it always succeeds. People, after all, are people.

The closest that Hex manages to come to genuine science is Phocian the Touched's lengthy and meticulous investigation of Antigonus's theory of the trotting horse. We hope that you have heard of neither of these gentlemen, since, to the best of our knowledge, they never existed. But then, neither did the Crab Civilisation – which didn't stop the crabs making their Great Leap Sideways. Our story here is modelled on real events, but we've simplified various otherwise distracting issues. With which we shall now distract you.

The prototype for Antigonus is the Greek philosopher Aristotle, a very great man who was even less of a scientist than Archimedes, whatever anyone has told you. In his *De Incessu Animalium* (*On the Gait of Animals*) Aristotle says that a horse cannot bound. The bound is a four-legged gait in which both front legs move together, then both back legs move together. He's right, horses don't bound. But that is the least interesting thing here. Aristotle explains *why* a horse can't bound:

> If they moved the fore legs at the same time and first, their progression would be interrupted or they would even stumble forward ... For this reason, then, animals do not move separately with their front and back legs.

Forget the horse: many quadrupeds do bound, so his reasoning, such as it is, must be wrong. And a gallop is very close to a bound, except that the left and right legs move at very slightly different times. If the bound were impossible, then by the same token so should the gallop be. But horses gallop.

Oops.

You can see that all this is a bit too messy to make a good story, so in the interests of narrativium we have replaced Aristotle by Antigonus, and credited him with a very similar theory about a long-standing historical conundrum: does a trotting horse always have at least one hoof on the ground? (In a trot, diagonally opposite legs move together, and the pairs hit the ground alternately.) This is the kind of question that must have been discussed in ale-houses and public baths since well before the time of Aristotle, because it's *just* out of reach of the unaided human eye. The first definitive answer came in 1874 when Eadweard Muybridge (born Edward Muggeridge) used high-speed photography to show that sometimes a trotting horse has all four feet off the ground at once. The proportion of times this occurs depends on the speed of the horse, and can be more than Phocian's 20 per cent. It can also be zero, in a slow trot, which further complicates the science. Allegedly, Muybridge's photographs won Leland Stanford Jr, a former Governor of California, the tidy sum of $25,000 in a bet with Frederick MacCrellish.

But what interest us here is not the science of horse locomotion, fascinating as that may be. It is how a scientific mind would go about investigating it. And Phocian shows that the Greeks could have made a lot more progress than they did, if they'd thought like a scientist. There were no technological barriers to solving such problems; just mental and (especially) cultural ones. The Greeks could have invented the phonograph, but if they did, it left no trace. They could have invented a clock, and the Antikythera mechanism shows they had the technique, but it seems that they didn't.

The slaves' use of songs to keep time has its roots in later history. In 1604 Galileo Galilei used music as a way to determine short intervals of time in some of his experiments on mechanics. A trained musician can mentally subdivide a bar into 64 or 128 equal parts, and even untrained people can distinguish an interval of a hundredth of a second in a piece of music. The Greeks could have used Galileo's method if they'd thought of it, and advanced science by 2,000 years. And they could have invented innumerable Heath-Robinson gadgets to study a moving horse, if it had occurred to them. Why didn't they? Possibly because, like Phocian, they were too tightly focussed on specific issues.

Phocian's approach to the trotting horse looks pretty scientific. First he tries the direct method: he gets his slaves to observe the horse while it is trotting, and *see* whether it is ever completely off the ground. But the horse is moving too fast for human vision to provide a convincing answer. So then he goes for the indirect approach. He thinks about Antigonus's *theory*, and homes in on one particular step: if the horse is off the ground, then it ought to fall over. That step can be tested in its own right, though in a different situation: a horse slung from a rope. (This way of thinking is called 'experimental design'.) If the horse does *not* fall over, then the theory is wrong. But this experiment is inconclusive, and even if the theory is wrong the conclusions could still be right, so he refines the hypothesis and invents more elaborate apparatus.*

We don't want to go too deeply into details of design here. We can think of ways to make the experiment workable, but the discussion would be a bit technical. For example, it seems necessary to make the roll of cloth, the Endless Road, move at a speed that is non-zero, but is also different from the natural speed with which the horse would move if its feet were actually hitting solid ground.† You might care to think about that, and you might even decide that we're wrong. And you might even be right.

We also acknowledge that Phocian's final experiment is open to many objections. And because the hooves of a trotting horse hit the ground in pairs, it is actually necessary to halve the total length of the charcoal smears before comparing them with the length of the cloth.

---

* In this, he is acting *exactly* like a scientist. Especially if it's very expensive apparatus.

† Gait analysts do put horses on treadmills. However, the closest parallel to Phocian's experiment is the widespread use of soot-covered cylinders to record insect movements.

No matter, these are mere elaborations of what would otherwise be an entirely transparent story: you understand what we're getting at.

Taking all this into account, was Phocian a scientist?

No. Hex has bungled again, for despite Phocian's years of visibly 'scientific' activity, he falls down in two respects. One, open to dispute, is not his fault: he has no peers, no colleagues. There are no other 'scientists' for him to work with, or to criticise him. He's on his own and ahead of his time.* Just as there cannot be just one wizard, there cannot be just one scientist. Science has a social dimension.† The second reason, though, is decisive. He is mortified when his work proves that Antigonus, the great authority, is wrong.

Any genuine scientist would give their right arm to prove that the great authority is wrong.

That's how you make your reputation, and it's also the most important way to contribute to the scientific endeavour. Science is at its best when it changes people's minds. Very little of it does that, in part because our minds have been built by a culture that is pervaded by science anyway. If a scientist manages to spend 1 per cent of the time discovering things that are not what they expected, they are doing amazingly well. But boy, does that 1 per cent count for a lot.

This, then, is science. Questioning authority. Complicity between theory and experiment. And being within a community of like-minded people to question *your* work. Preferably accompanied by a conscious awareness of all of the above, and gratitude to your friends and colleagues for their criticisms. And what's the aim? To find timeless truths? No, that's asking too much. To stop frail humans from falling for

* There have been many others. One of our favourites is Sir George Cayley, the early nineteenth-century aeronautical pioneer. He did sterling work on wing design, invented the light-tension wheel (effectively the modern bicycle wheel) as a light wheel for aircraft, and would almost certainly have achieved powered flight if only anyone had got around to inventing the internal combustion engine. He didn't go mad, but he did experiment with an engine that ran on gunpowder.

† We're in danger of heading into postmodernism here, which is a very bad idea when discussing an ancient Greek, and even more so when he's fictitious. Suffice it to say that science also involves stringent reality-checks, and therefore is not a *purely* social activity.

plausible falsehoods? Yes – including those of people who at least look and sound just like you. And to protect people from their willingness to believe a good story, just because it sounds right and doesn't upset them. And to protect them from the firm smack of authority, too.

It took humanity a long time to arrive at the scientific method. No doubt the reason for the delay was that if you do science properly, you often find yourself overturning entrenched, well-established beliefs, including your own entrenched, well-established beliefs. Science is not a belief system, but many areas of human activity are, so it is not surprising to find that the early developers of science often found themselves in conflict with authority. Perhaps the best-known example of this is Galileo, who ran into trouble with the Inquisition because of his theories about the solar system. Sometimes science exposes you to the firm smack instead.

Science, then, is not just a body of teachable facts and techniques. It is a way of thinking. In science, established 'facts' are always open to question,* but few scientists will listen to you unless you can offer some evidence that the old ideas are wrong. If the people who invented those ideas are dead, then alternatives can quickly gain acceptance, and the scientific method is working well. If the people who invented those ideas are still around, in influential positions, then they can put a lot of obstacles in the way of the new suggestion and the people who proposed it. Then science is working badly, because people are behaving like people. Even so, the new idea still can displace the accepted wisdom. It just takes longer and needs really solid evidence.

Let's contrast science with alternative ways of thinking about the universe. The Discworld worldview is that the universe is run by

---

* Some current controversies, all 'respectable' – that is, with serious evidence for both sides – include: Is new variant CJD related to BSE (mad cow disease)? Has the human sperm count fallen? Was the Moon formed by a Mars-sized body hitting the Earth? Will the universe ever stop expanding? How are birds related to dinosaurs? Is quantum mechanics really random? Was there ever life on Mars? Is the triple-alpha process evidence that our universe is special? And is there anything that does *not* contain nuts?

magic: things happen because people want them to happen. You still have to find the right spell, or the narrative imperative has to be so strong that those things will happen anyway even if people *don't* want them to, but the universe exists in order to be there for people.

On Discworld and Roundworld, the worldview of the priesthood is similar, but with one important difference. They believe that the universe is run by gods (or a god): things happen because the gods want them to happen, don't care if they happen, or have some ineffable long-term aim in view. However, it is possible for people to ask the priests to intercede with the gods, on their behalf, in the hope of influencing the gods' decisions, at least in minor ways.

The philosophical worldview, exemplified by Antigonus, is that the nature of the world can be deduced by pure thought, on the basis of a few deep, general principles. Observation and experiment are secondary to verbal reasoning and logic.

The scientific worldview is that what people want has very little to do with what actually happens, and that it is unnecessary to invoke gods at all. Thought is useful, but empirical observations are the main test of any hypothesis. The role of science is to help us find out how the universe works. Why it works, or what manner of Being ultimately controls it, if any, is not a question that science is interested in. It is not a question to which anyone can give a testable answer.

Oddly enough, this hands-off approach to the universe has given us far more control over it than magic, religion or philosophy have done. On Roundworld, magic doesn't work, so it offers no control at all. Some people believe that prayer can influence their god, and that in this way human beings can have some influence over the world in which they live, like a courtier at a king's ear. Other people have no such beliefs, and consider the role of prayer to be largely psychological. It can have an effect on *people*, but not on the universe itself. And philosophy has a tendency to follow rather than lead.

Science is a form of narrativium. In fact, all four approaches to the universe – magic, religion, philosophy and science – involve the construction of stories about the world. Oddly enough, these different kinds of story often have many parallels. There is a distinct resemblance between many religious creation myths and the cosmologists' 'Big Bang' theory of the origin of the universe. And the monotheistic idea that there is only one God, who created everything and runs everything, is suspiciously close to the modern physicists' idea that there should be a single Theory of Everything, a single fundamental physical principle that unites both relativity and quantum mechanics into a satisfying and elegant mathematical structure.

The act of telling stories about the universe may well have been more important to the early development of humanity, and for the initial growth of science, than the actual content of the stories themselves. Accurate content was a later criterion. When we start telling stories about the universe, the possibility arises of comparing those stories with the universe itself, and refining how well the stories fit what we actually see. And that is already very close to the scientific method.

Humanity seems to have started from a rather Discworldly view, in which the world was inhabited by unicorns and werewolves and gods and monsters, and the stories were used not so much to explain how the world worked, but to form a crucial part of the cultural Make-a-Human kit. Unicorns, werewolves, elves, fairies, angels, and other supernatural were not real. But that didn't actually matter very much: there is no problem in using unreal things to programme human minds.* Think of all those talking animals.

The models employed by science are very similar in many respects. They, too, do not correspond exactly to reality. Think of the old model of an atom as a kind of miniature solar system, in which tiny hard particles called electrons whirl around a central nucleus consisting of other kinds of tiny hard particles: protons and neutrons. The atom is not 'really' like that. But many scientists still use this picture today as the basis for their investigations. Whether this makes sense depends upon what problem they are working on, and when it doesn't make sense, they use something more sophisticated, like the description of an atom as a probable cloud of 'orbitals' which represent not electrons, but places where electrons could be. That model is more sophisticated, and

---

* Yes, in some cases, it is claimed, werewolves and vampires have their roots in rare human medical conditions. Now try angels and unicorns …

it fits reality more closely than a mini solar system, but it still isn't 'true'.

Science's models are not *true*, and that's exactly what makes them useful. They tell simple stories that our minds can grasp. They are lies-to-children, simplified teaching stories, and none the worse for that. The progress of science consists of telling ever more convincing lies to ever more sophisticated children.

Whether our worldview is magical, religious, philosophical or scientific, we try to alter the universe so that we can convince ourselves that we're in charge of it. If our worldview is magical, we believe that the universe responds to what we want it to do. So control is just a matter of finding the right way to instruct the universe about what our wishes are: the right spell. If our worldview is religious, we know that the gods are really in charge, but we hold out the hope that we can influence their decisions and still get what we want (or influence ourselves to accept whatever happens …). If our worldview is philosophical, we seldom tinker with the universe ourselves, but we hope to influence how others tinker. And if our worldview is scientific, we start with the idea that controlling the universe is not the main objective. The main objective is to *understand* the universe.

The search for understanding leads us to construct stories that map out limited parts of the future. It turns out that this approach works best if the map does not foretell the future like a clairvoyant, predicting that certain things will happen on certain days or in certain years. Instead, it should predict that if we do certain things, and set up a particular experiment in particular circumstances, then certain things should happen. Then we can do an experiment, and check the reasoning. Paradoxically, we learn most when the experiment fails.

This process of questioning the conventional wisdom, and modifying it whenever it seems not to work, can't go on indefinitely. Or can it? And if it stops, when does it stop?

Scientists are used to constant change, but most changes are small: they refine our understanding without really challenging anything. We take a brick out of the wall of the scientific edifice, polish it a bit, and put it back. But every so often, it looks as if the edifice is actually finished. Worthwhile new questions don't seem to exist, and all attempts to shoot down the accepted theory have failed. Then that area of science becomes established (though still not 'true'), and nobody wastes their time trying to change it any more. There are always other sexier and more exciting areas to work on.

Which is much like putting a big plug in a volcano. Eventually, as the pressure builds up, it will give way. And when it does, there will be a very big explosion. Ash rains down a hundred miles away, half the mountain slides into the sea, everything is altered …

But this happens only after a long period of apparent stability, and only after a huge fight to preserve the conventional ways of thinking. What we then see is a paradigm shift, a huge change in thought patterns; examples include Darwin's theory of evolution and Einstein's theory of relativity.

Changes in scientific understanding force changes in our culture. Science affects how we think about the world, and it leads to new technologies that change how we live (and, when misunderstood, deliberately or otherwise, some nasty social theories, too).

Today we *expect* big changes during our lifetimes. If children are asked to forecast the future, they'll probably come up with science-fictional scenarios of some kind – flying cars, holidays on Mars, better and smaller technology. They are probably wrong, but that doesn't matter. What matters is that today's children do not say: 'Change? Oh, everything will probably be pretty much the same. I'll be doing just the same things that my Mum and Dad do now, and their Mum and Dad did before them.' Whereas even fifty years ago, *one grandfather*, that was generally the prevailing attitude. Ten or eleven grandfathers ago, a big change for most people meant using a different sort of plough.

And yet … Underneath these changes, people are still people. The basic human wants and needs are much as they were a hundred grandfathers ago, even if we ever do take holidays on Mars (all that beach …). The realisation of those needs may be different – a hamburger instead of a rabbit brought down with an arrow you made yourself – but we still want food. And companionship and sex and love and security and lots of other familiar things.

The biggest significant change, one that really does alter what it is like to be human, may well be modern communication and transportation.

The old geographical barriers that kept separate cultures separate have become almost irrelevant. Cultures are merging and reforming into a global multiculture. It's hard to predict what it will look like, because this is an emergent process and it hasn't finished emerging yet. It may be something quite different from the giant US shopping mall that is generally envisaged. That's what makes today's world so fascinating – and so dangerous.

Ultimately, the idea that we are controlling our universe is an illusion. All we know is a relatively small number of tricks, plus one great generic trick for generating more small tricks. That generic trick is the scientific method. It pays off.

We have also the trick of telling stories that work. By this stage in our evolution, we are spending most of our lives in them. 'Real life' – that is, the real life for most of us, with its MOT tests and paper wealth and social systems – is a fantasy that we all buy into, and it works precisely because we all buy into it.

Poor old Phocian tried hard, but found that the old stories weren't true when he hadn't quite got as far as constructing a new one. He performed a reality check, and found that there wasn't one – at least, not one he'd like to believe was real. He suddenly saw a universe with no map. We've got quite good at mapping, since then.

TWENTY-THREE

PARAGON OF
ANIMALS

 THE WIZARDS WENT BACK TO DEE'S HOUSE in sombre mood, and spent the rest of the week sitting around and getting on one another's nerves. In ways they couldn't quite articulate, they'd been upset by the story.

'Science is dangerous,' said Ridcully at last. 'We'll leave it alone.'

'I think it's like with wizards,' said the Dean, relieved to be having a conversation again. 'You need to have more than one of them, otherwise they get funny ideas.'

'True, old friend,' said Ridcully, probably for the first time in his life. 'So … science is not for us. We'll rely on common sense to see us through.'

'That's right,' said the Lecturer in Recent Runes. 'Who cares about trotting horses anyway? If they fall over they've only got themselves to blame.'

'As a basis for our discussion,' said Ridcully, 'let us agree on what we have discovered so far, shall we?'

'Yes. It's that whatever we do, the elves always win,' said the Dean.

'Er … I know this may sound stupid …' Rincewind began.

'Yes. It probably will,' said the Dean. 'You haven't been doing very much since we got back, have you?'

'Well, not really,' said Rincewind. 'Just walking around, you know. Looking at things.'

'Exactly! You haven't read a single book, am I correct? What good is walking around?'

'Well, you get exercise,' said Rincewind. 'And you notice things. Yesterday the Librarian and I went to the theatre ...'

They'd got the cheapest ticket, but the Librarian paid for two bags of nuts.

They'd found, once they had settled into this period, that there was no point in trying to disguise the Librarian too heavily. With a jerkin, a big floppy hood and a false beard he looked, on the whole, an improvement on most of the people in the cheap seats, the cheap seats in this case being so cheap they consisted, in fact, of standing up. The cheap feets, in fact.

The play had been called *The Hunchback King*, by Arthur J. Nightingale. It hadn't been very good. In fact, Rincewind had never seen a worse-written play. The Librarian had amused himself throughout by surreptitiously bouncing nuts off the king's fake hump. But people had watched it in rapt fascination, especially the scene where the king was addressing his nobles and uttered the memorable line: 'Now is the December of our discontent – *I want whichever bastard is doing that to stop right now!*'

A bad play but a good audience, Rincewind mused after they had been thrown out. Oh, the play was a vast improvement on anything the Shell Midden Folk could have dreamed up, which would have to be called 'If We'd Invented Paint We Could Watch It Dry', but the lines sounded wrong and the whole thing was laboured and had no flow. Nevertheless, the faces of the watchers had been locked on the stage.

On a thought, Rincewind had put a hand over one eye and, concentrating fearfully, surveyed the theatre. The one available eye watered considerably but *had* revealed, up in the expensive seats, several elves.

They liked plays, too. Obviously. They wanted people to be imaginative. They'd given people so much imagination that it was constantly hungry. It would even consume the plays of Mr Nightingale.

Imagination created monsters. It made you afraid of the dark, but not of the dark's real dangers. It peopled the night with terrors of its own.

So, therefore ...

Rincewind had an idea.

'I think we should stop trying to influence the philosophers and schol-ars,' he said. 'People with minds like that believe all sort of things all the time. You can't stop them. And science is just too *weird*. I keep thinking of that poor man—'

'Yes, yes, yes, we've been through all this,' said Ridcully wearily. 'Get to the *point*, Rincewind. What have you got to say that's new?'

'We could try teaching people art,' said Rincewind.

'Art?' said the Dean. 'Art's for slackers! That'd make things *worse!*'

'Painting and sculpture and theatre,' Rincewind went on. 'I don't think we should try to stop what the elves began. I think we ought to encourage it as much as possible. Help the people here to get really good at imagining things. They're not quite there yet.'

'But that's just what the elves want, man!' snapped Ridcully.

'Yes!' said Rincewind, almost drunk with the novelty of having an idea that didn't include running away. 'Let's *help* the elves! Let's help them to destroy themselves.'

The wizards sat in silence. Then Ridcully said: 'What are you talk-ing about?'

'At the theatre I saw lots of people who wanted to believe that the world is different from the reality they see around them,' said Rincewind. 'We could—' He sought a way into Ridcully's famously hard-to-open mind. 'Well, you know the Bursar?' he said.

'A gentlemen of whose existence I am aware on a daily basis,' said Ridcully gravely. 'And I'm only glad that this time we've left him with his aunt.'

'And you remember how we cured his insanity?'

'We didn't cure it,' said Ridcully. 'We just doctored his medicine so that he permanently *hallucinates* that he is sane.'

'Exactly! You use the disease as the cure, sir! We made him *more* insane, so now he's sane again. Mostly. Apart from the bouts of weight-lessness, and, er, that business with the—'

'Yes, yes, all right,' said Ridcully. 'But I'm still waiting for the *point* of this.'

'Are you talking about fighting like those monks up near the Hub?' said the Lecturer in Recent Runes. 'Skinny little chaps who can throw big men through the air?'

'Something like that, sir,' said Rincewind.

Ridcully prodded Ponder Stibbons.

'Did I miss a bit of conversation there?' he said.

'I think Rincewind means that if we take the elves' work even further it'll somehow end up defeating them,' said Ponder.

'Could that work?'

'Archchancellor, I can't think of anything better,' said Ponder. 'Belief doesn't have the same power on this world as it does on ours, but it is still pretty strong. Even so, the elves *are* here. They are a fixture.'

'But we know they ... sort of *feed* on people,' said Rincewind. 'We want them to go away. Um ... and I've got a plan.'

'*You* have a plan,' said Ridcully, in a hollow voice. 'Does anyone *else* have a plan? Anyone? Anyone? Someone?'

There was no reply.

'The play I saw was awful,' said Rincewind. 'These people might be a lot more creative than the Shell Midden People, but they've still got a long way to go. My plan ... well, I want us to move this world into the path of history that contains someone called William Shakespeare. And absolutely does not contain Arthur J. Nightingale.'

'Who's Shakespeare?' said Ponder.

'The man,' said Rincewind, 'who wrote this.' He pushed a battered manuscript across the table. 'Read it out from where I've marked it, will you?'

Ponder adjusted his spectacles, and cleared his throat.

'What a piece of work is, er, this is awful handwriting ...'

'Let me,' said Ridcully, taking the pages. 'You don't have the voice for this sort of thing, Stibbons.' He glared at the paper, and then: 'What a piece of work is a man! How noble in reason ... how infinite in faculty ... in form and moving, how express and admirable! In action, how like an angel! In apprehension, how like a god! The beauty of the world, the paragon of animals ...'

He stopped.

'And this man lives *here*?' he said.

'Potentially,' said Rincewind.

'This man stood knee deep in muck in a city with heads on spikes and wrote *this*?'

Rincewind beamed. 'Yes! In his world, he is probably the most influential playwright in the history of the species! Despite requiring a lot of tactful editing by most directors, because he had his bad days just like everyone else!'

'By "his world" you mean—?'

'Alternate worlds,' muttered Ponder, who was sulking. He'd once played the part of Third Goblin in a school play and felt that he had rather a good speaking voice.

'You mean he should be here but ain't?' Ridcully demanded.

'I think he should be here but can't be,' said Rincewind. 'Look, these aren't the Shell Midden people, it's true, but artistically they're pretty low down the scale. Their theatre is awful, they haven't got any decent artists, they can't carve a decent statue – this world isn't what it should be.'

'And?' said Ponder, still smarting.

Rincewind signalled to the Librarian, who ambled around the table handing out small, green, cloth-bound books.

'This is *another* play he will write … is … writing … wrote … will have written,' he said. 'I think you'll agree that it could be very important …'

The wizards read it. They read it again. They had a huge argument, but there was nothing unusual about that.

'It's an astonishin' play, in the circumstances,' said Ridcully, eventually. 'And some of it is a bit familiar!'

'Yes,' said Rincewind. 'And I think that's because he'll write it after listening to you. We need him to. This is a man who can tell the audience, *tell* the audience that they're watching a bunch of actors on a tiny stage and *then* make them see a huge battle, right there in front of their eyes.'

'Did I miss that bit?' said the Lecturer in Recent Runes, leafing hurriedly through the pages.

'That's in another play, Runes,' said Ridcully. 'Do try to keep up. Well, Rincewind? Let's assume, shall we, that we're going along with your plan? We have to make sure this man exists here and writes this play in this world, do we? *Why?*'

'Can I leave that to Stage Two, sir? It will become obvious, I hope, but you never know if there are elves listening.'

The wizards were automatically impressed by the idea that this was a two-stage plan, but Ridcully persisted: 'I put it to you, Rincewind, that this is exactly the kind of play elves would *want* him to write.'

'Yes, sir. That's because they're stupid. Not like you, sir.'

'We have Hex's computational power,' said Ponder. 'It should be possible to make sure he turns up in this world, I think.'

'Um ... yes,' said Rincewind. 'But first we have to make the world the kind he *can* turn up in. This may take a bit of work. Some travelling may be involved. Back in time ... for thousands of years ...'

Firelight glowed off the cave walls. The wizards sat on one side of the fire, on the big rock ledge overlooking the scrubland. The Stinky Cave People sat on the other.

The cave people watched the wizards with something like awe, but only because they'd never seen people eat like that. It was Ridcully who'd suggested that people bearing huge amounts of food are welcome practically anywhere, but the other wizards considered that this was just an excuse for him to make a crude but serviceable bow and go and happily slaughter quite a lot of wildlife.

The wildlife was mostly leftovers now. The wizards moaned about the lack of onions, salt, pepper, garlic and, in Rincewind's case, potatoes, but there was certainly no lack of meat.

They'd spent two weeks doing this, in caves across the continent. They were getting used to it, although bowel movements were becoming a problem.

Rincewind, however, was sitting some way from the fire with Burnt Stick Man.

Being good at languages was, here, not such an important skill as simply making yourself understood. But Burnt Stick Man was a quick study, and Rincewind already had several weeks of practice. While the dialogue took place in inflections and emphasis based upon the syllable 'grunt' aided by gestures, the translation went like this:

'Okay, so you've mastered the idea of charcoal, but may I draw your attention to these pigments I have here? They're Whiiite, very simple, Redddd, like blood, and Yell-low, like, er, egg yolks. Cluck cluck aaargh cackle? And this fourth colour is some sickly brown ochre I found which we'll call for the moment "baby poo".'

'With you so far, Pointy Hat Man.' This was conveyed by an enthusiastic nod.

'So here's the big tip. Not many people know this,' said Rincewind. 'You take your animals, right, which you've already been trying to draw, well done, but you what we call "colour" them. You have to work hard on this bit. A chewed piece of wood will be your friend here. See how by a careful mixture of tints I'm giving it a certain, oh, *je ne sais quoi?*'

'Hey, that looks like a *real* buffalo! Scary stuff!'

'It gets better. May I have the charcoal? Thank you. What's this?' Rincewind carefully drew another figure.

'Man with big [expressive gesture]?' said Burnt Stick Man.

'What? Oh. Sorry, I got that wrong … I mean *this* …'

'Man with spear! Hey, he's throwing it at the buffalo!'

Rincewind smiled. There had been a few false starts over the last couple of weeks, but Burnt Stick Man had exactly the right sort of mind. He was impressively simple, and people with truly simple minds were very rare.

'I knew there was something intelligent about you the moment I saw you,' he lied. 'Maybe it was the way your brow ridge came around the corner only two seconds before the rest of you did.' Burnt Stick Man beamed. Rincewind went on: 'And the question you've got to ask yourself now is: how *real* is this picture, really? And where was the picture before I drew it? What is going to happen now it's on the wall?'

The wizards watched from the circle of firelight.

'Why's the man poking at the picture?' said the Dean.

'I think he's learnin' about the power of symbols,' said Ridcully. 'Hey, if anyone doesn't want any more ribs I'll finish 'em.'

'No barbecue sauce,' moaned the Lecturer in Recent Runes. 'How long before there's an agricultural revolution?'

'Could be a hundred thousand years, sir,' said Ponder. 'Perhaps a lot more.' The Lecturer in Recent Runes groaned and put his head in his hands.

Rincewind came and sat down. The rest of Burnt Stick Man's clan, greasy to the eyebrows with free food, watched him cautiously.

'That seemed to go well,' he said. 'He's definitely working out the link between pictures in his head and real life. Any potatoes yet?'

'Not for thousands of years,' groaned the Lecturer in Recent Runes.

'Damn. I mean, here's meat. There should be potatoes. How hard is that for a world to understand? Vegetables are less *complicated* than meat!' He sighed, and then stared.

Burnt Stick Man, who had been staring motionless at the drawing for a while, ambled to another rock wall and picked up a spear. He squinted at the buffalo drawing, which did indeed seem to move as the firelight flickered, paused, and then hurled the spear at it and ducked behind a rock.

'Gentlemen, we've found our genius and we're on our way,' said Rincewind. 'Ponder, can Hex move some buffaloes to right outside this cave at dawn tomorrow?'

'That shouldn't be hard, yes.'

'Good.' Rincewind looked around. 'And there's quite a few tall trees here, too. Which is just as well.'

It was dawn, and the tree was full of wizards.

The ground below was full of buffalo. Hex had moved an entire herd, which was now more or less penned in amongst the rocks and trees.

And, on the rocky ledge in front of the bewildered, panicking creatures, Burnt Stick Man and the other hunters stared down in disbelief.

But only for a moment. They had spears, after all. They got two of the creatures before the rest thundered away. And, afterwards, people were certainly showing Burnt Stick Man a bit of respect.

'All right, I think I see what you're getting at,' said Ridcully, as the wizards very carefully climbed down.

'Well, *I* don't,' said the Dean. 'You're teaching them basic magic. And that doesn't work here!'

'They *think* it does,' said Rincewind.

'But that was only because we helped them! What're they going to do tomorrow when he does another painting and no buffaloes turn up?'

'They'll think it's experimental error,' said Rincewind. 'Because it's so *sensible*, isn't it? You draw a magic picture, and the real thing turns up! It's *so* sensible that they'll take a lot of convincing that it doesn't work. Besides ...'

'Besides what?' said Ponder.

'Oh, I was thinking that if Burnt Stick Man is *really* sensible he'll keep an eye on the movements of the local animals and make sure he paints his pictures at the right time.'

Some more weeks went by. There were lots of men like Burnt Stick Man.

And even Red Hands Man ...

'... so,' said Rincewind, as he sat by the river, squeezing the clay, 'it's quite easy to make *other* things out of it than snakes.'

'Snakes are easy,' said Red Hands Man, stained with ochre to the armpits.

'And there's lots of snakes around here, is there?' said Rincewind. It looked like prime snake country.

'Lots of them.'

'Ever wondered why? You play around rolling snakes out of clay, and snakes turn up?'

'*I'm* making the snakes?' said Red Hands Man. 'How can that be? I was only doing it because of the enjoyable tactile sensations!'

'It's an intriguing thought, isn't it?' said Rincewind. 'But it's okay, I won't tell anyone else.'

Red Hands Man stared at his hands as if examining two lethal instruments. He seemed a little less bright than Burnt Stick Man.

'Ever thought about making something else?' said Rincewind. 'Something more edible?'

'Fish are good to eat,' Red Hands Man conceded.

'Why not try making a clay fish?' said Rincewind, with a sincere smile.

Next morning, it rained trout.

In the afternoon a very happy Red Hands Man, now hailed as the saviour of the clan that lived among the reeds, made a model of a big fat woman out of clay.

The wizards discussed the moral implications of allowing Hex to rain enormous women over a wide area. The debate took a long time, with many pauses for inward reflection, but at last the Dean was voted down. It was agreed that if you gave a man a fat woman, he'd just have a fat woman for a day, but if you helped a man become a very important man because he had the secret of buffaloes or fish, he could get himself as many fat women as he wanted.

Next morning they went forward a thousand years in time. There was hardly an unadorned cave on the continent, and quite a lot of fat women.

They went further ...

In a forest clearing, a man was making a god out of wood. Either it wasn't a very good carving, or it was a good carving but an ugly god.

The wizards watched.

And the Queen of the Elves appeared, with a couple of elves in attendance. They were male or, at least, appeared male. The queen was angry.

'What are you doing, wizards?' she snapped.

Ridcully gave her a nod of annoying friendliness. 'Oh, just a little ... what are we calling it, Stibbons?'

'A sociological experiment, Archchancellor,' said Ponder.

'But you've been teaching them art! And sculpture!'

'And music,' said Ridcully happily. 'The Lecturer in Recent Runes is rather good with a lute, it turns out.'

'Only in a very amateur way, I'm afraid,' said the Lecturer in Recent Runes, blushing.

'Dashed easy to make, a lute,' said Ridcully. 'You just need a tortoise shell and some sinews and you're well away. I myself have been renewing my acquaintance with the penny whistle of my boyhood, although I fear that the Dean's expertise with the comb-and-paper leaves something to be desired ...'

'And *why* are you doing all this?' the queen demanded.

'Are you angry? We thought you'd be pleased,' said Ridcully. 'We thought you *wanted* them this way. You know – imaginative.'

'*He* created music?' said the Queen, glaring at the Lecturer in Recent Runes, who gave her an embarrassed wave.

'Oh, no, I assure you,' he said. 'Er, they'd worked up to, you know, basic percussion, the conch shell and so on, but it was all rather dull. We just helped them along a bit.'

'Gave them a few tips,' said Ridcully, jovially.

The Queen's eyes narrowed. 'Then you are *planning* something!' she said.

'Aren't they doing well?' said Ridcully. 'Look at that chap over there. *Visualisin'* a god. One with woodworm and knotholes, but pretty good all the same. Quite complex mental processes, really. We thought that if you want people with wild imaginations, then we'd help them to be really good at it. They'll fill the world with dragons and gods and monsters for you. You *want* that.'

The Queen gave him another look, and it was the look of a person with no sense of humour who nevertheless suspects that there's some joke somewhere that is on them.

'Why should you help us?' she said. 'You told me to consume your underthings!'

'Well, it's not as though this world is important enough to fight over,' said Ridcully.

'One of you isn't here,' said the Queen. 'Where is the stupid one?'

'Rincewind?' said the Archchancellor, with an innocent air that would not have fooled any human for a moment. 'Oh, he's doing pretty much the same thing, you know. Helping people imagine things. Which, I think, is what you want.'

TWENTY-FOUR

THE EXTENDED PRESENT

ART? IT LOOKS SUPERFLUOUS. Few of the stories we tell about human evolution, the *Homo sapiens* bit, see music or art as being integral to the process. Oh, it often comes in as a kind of epiphenomenon, as evidence of how far we'd got: 'Just look at those wonderful cave paintings, statuettes, polished jewellery and ornaments! That shows that our brain was bigger/better/more loving/nearer to that of the Lecturer in Recent Runes ...' But art has not been portrayed as a necessary part of the evolution that made us what we are; nor has music.

So why are Burnt Stick Man and Red Hands Man dabbling in art, and why does Rincewind want to encourage them?

We've been told the story of The Naked Ape doing sex, we've had Gossiping Apes and Privileged Apes, various kinds of apes becoming intelligent on the seashore or running down gazelles on the savannah. We've had lots of development-of-intelligence stories culminating in Einstein; we have given you the privilege/puberty ritual/selection story that culminates in Eichmann and Obedience to Authority; but we have not presented a version of our evolution whose culmination is Fats Waller, Wolfgang Amadeus Mozart, or even Richard Feynman on the bongo drums.

Well, now we will.

Music is an important part of most people's lives, and this is continually reinforced by film and television. Background music is constantly informing us of imminent screen events, of tension and release, of characters' thoughts and, particularly, of their emotional states. It is very difficult for anyone brought up in the muzak environment of the twentieth century to imagine what the 'primitive' state of human musical sense can have been.

When we listen to the music of far peoples, of 'primitive' tribes, we have to appreciate that their music has had as long to develop as Beethoven, and much longer than jazz. Like the amoeba or the chimpanzee, their music is contemporary with us, not ancestral, though it sounds primitive, just as they *look* primitive. And we wonder whether we are listening for the right things in the right way. It is tempting to think that popular music, going for instant appeal, might illuminate whatever inner structure of our brains 'fits', and is satisfied by, a musical theme. If we were orthodox geneticists, we might have said 'genes for music' there. But we didn't.

In recent years, neuroscientists have developed techniques that allow us to look at what brains do when we carry out various actions. In particular, they reveal which bits of brain are active when we enjoy music. At the moment, with the terribly poor spatial and temporal resolution that we get from MRI and PET scans, all we can see is that music excites the right side of the brain. If we are familiar with the music, then the brain's memory-regions turn on, and if we analyse it or try to pick up the lyrics, then the verbal-analysis parts light up. And opera picks up both of them, which could be why Jack likes it: he enjoys having his brain put through a blender.

Our affinity with music starts early. In fact, there's a lot of evidence that if we hear music in the womb, then it can affect our later musical preferences. Psychologists play music to babies as soon as they start kicking, and have discovered that they can categorise it, like we adults do, and into the same categories. If we play them Mozart, they stop kicking for a bit, about fifteen minutes; then they start kicking again, perhaps with some relation to the rhythm. The evidence is claimed, but it isn't very persuasive. If we then continue with a different bit of Mozart, or Haydn or Beethoven, then the kicking pauses, but it resumes after a minute or so. The Beatles, Stravinsky, sacred chants, or New Orleans jazz, make them pause for much longer, ten minutes or so.

Playing the same pieces months later reveals that the baby has some memory of the style as well as of the instruments. Apparently, a quartet by Mozart triggers recognition of the 'Mozart' style just as effectively as a Mozart symphony. Our brains have sophisticated music-recognition modules, and we can use them before we speak, indeed before we are born. Why?

We're looking for the essence of music – as if we *knew* what the essence of sex was for the Naked Ape, or the essence of obedience for Eichmann – or come to that, what it means to be the most intelligent/extelligent creature on Roundworld. What we want is a story that puts the arts, and music, into an explanation of How We Got Here, and why we waste all that money on the arts faculties of universities. Why is Rincewind so keen to bring art and music to our ancestors?

It was very common in the early years of the twentieth century to copy the music of 'primitive' tribes. Examples include Stravinsky's *Rite of Spring* and Manuel de Falla's *Fire Dance*, where the musical style was thought to give a primitive authenticity. People thought that Bronislaw Malinowski's tales of the Trobriand Islanders, with their amazing lack of the civilised sexual repressions so publicised by Freud in Viennese society, showed that Natural Humans were happier and less corrupt, and that their music – for flutes and drums – conveyed their state of innocence more effectively than classical symphonies. Jazz, invented by supposedly 'primitive' black musicians down in New Orleans, had resonances that seemed natural, animal (and, for certain Christians, evil). It was almost as if music were a language, parallel to the words, developed in different societies with different emphases, and more revealing of the nature of the people than other aspects of their culture.

This is the way the media have played it, and like the Flintstones and Stone Age society, we have an overlay of this outlook that it's very difficult to get away from. Margaret Mead, who was taken for a ride by her native girl friend and told the resulting story in *Coming of Age in Samoa: A Psychological Study of Primitive Youth for Western Civilization*, romanticised their music and dances in exactly this way.

When Hollywood needs to show the primitive-but-spiritual nature of Indian braves, cannibal tribes in Borneo, or Hawaiian indigenes, it shows us the rain dance, the marriage music, and the hula girls. When we go to these places, the locals put on these dances for us because it brings in tourist money. The complicity between muzak, hula dances, opera and background music in Hollywood films has completely buried our abilities to sort out what constitutes 'natural' art or music.

However, that's not what we want anyway. 'Natural' is an illusion. Desmond Morris made a lot of money selling paintings done by apes. The apes clearly enjoyed the whole business, and so did Morris, and presumably so did the people who bought them and looked at them in art galleries. There is also an elephant that paints, and signs its paintings. Sort of. There's a segment of modern painting whose philosophy seems to relate to this quest for the genuinely primitive. One side is the tackiest, painting by children, which clearly demonstrates the stepwise effects of the culture – the extelligence – on their burgeoning intelligence. To our inexpert eyes, though, these paintings demonstrate only the enormous gratification achieved by some parents in response to minimal effort by their children.

Another aspect, more intellectual, is the move towards apparently real-world constraints, like cubism, or attempts to develop styles that force us to re-evaluate how we see, like Picasso's profile faces but with the two eyes on one side. There is a very common modern form that arranges rectangles of paper with different textures, or sprays sparse paint droplets according to some minimal rule, or scatters charcoal dust on a bold swirly bright oil-paint background and then combs it into the texture and pattern of the whole canvas. All of these can give pleasure to the eye. Why? How do they differ from natural objects, some of which also give considerable pleasure?

Now we want to make a giant leap and bring Mozart, jazz, paper-texture and charcoal-swirl oil paintings into the same frame. We think that this frame naturally includes ancient cave-paintings, which we know to be early, so have more claim to being genuinely primitive, if we could only look at them with the eyes and minds of viewers contemporary with the artist. The same problem occurs with Shakespeare, too: we no longer have the ears or minds – the extelligence – of the first Elizabethan age.

We have to be more than a bit scientific here. We have to consider how we perceive light, sound, touch – what our sense organs tell us. For a start, they don't, and this is the first lesson. In his book *Consciousness Explained*, Daniel Dennett is very critical of the Cartesian Theatre* picture of consciousness. In this picture, we imagine ourselves sitting in a little theatre in our minds, where our eyes and ears pipe in pictures and sounds from the outside world. In school we all learned that the eye is like a camera, and that a picture of the world is imaged in the plane of the retina, as if that was the difficult bit. No, the difficult bit *starts* there, with different elements of that picture taking different routes into different parts of the brain.

When you see a moving red bus, the features 'moving', 'red' and 'bus' are separated fairly early in the brain's analysis of the scene … and they don't just get put together again to synthesise your mental picture. Instead, your picture is synthesised from lots of clues, lots of bits, and nearly all of what you 'see' as you look around the room is only 'there' in your brain. It's not at all like a TV picture. It is not picked up instantly and updated, but nearly all of that 'detailed' surround is invented as a kind of wallpaper around the little bit that has your attention. Most of the details are not present *as such* in your mind at all, but that's the illusion that your mind presents to you.

When we see a painting … except, again, we don't. There are several ways to convince people that they invent what they 'see', that perception is not simply a copy of the eye's image on the retina. There is, for example, a blind spot on the retina where the optic nerve leaves it. This is big. It's as big as 150 full moons (that's not a misprint: a hundred and fifty). Not that the moon is as big, to our eyes, as we usually think – and certainly not as big as Hollywood repeatedly shows it. We 'see' the full moon as much bigger than it 'is' (sorry, we have to use some trick to separate what's in your mind from reality out there),

---

* Cartesian, again, because of Descartes, whose *cogito ergo sum* and mind-is-different-stuff-from-matter still influence pop philosophy.

especially when it's near the horizon. The best way to appreciate that is to demonstrate to yourself that the moon's image is the size of your little fingernail at arm's length. Hold out your arm, and the tip of your littlest finger more than covers the moon. So the blind spot is smaller than our description may have suggested, but it's still a big chunk of the retinal image. We don't notice any hole in the picture we get of the outside world, though, because the brain fills in its best estimate of what's missing.

How does the brain *know* what's missing from right in front? It doesn't, and it doesn't have to: that's the point. Although 'fills in' and 'missing' are traditional terms in this area of science, they are, again, misleading. The brain doesn't *notice* that anything is missing, so there isn't a gap to be filled in. The neurons of the visual cortex, the part of the brain that analyses that retinal image into a scene that we can recognise and label, are wired up in elaborate ways, which reinforce certain perceptual prejudices.

For example, experiments with dyes that respond to the brain's electrical signals show that the first layer of the visual cortex detects lines – edges, mostly. The neurons are arranged in local patches, 'hypercolumns', which are assemblies of cells that respond to edges aligned along about eight different directions. Within a hypercolumn, all connections are inhibitory, meaning that if one neuron thinks it has seen an edge pointing along the direction to which it is sensitive, then it tries to stop the other neurons from registering anything at all. The result is that the direction of the edge is determined by a majority vote. In addition, there are also long-range connections between hypercolumns. These are excitatory, and their effect is to bias neighbouring hypercolumns to perceive the natural continuation of that edge, even if the signal they receive is too weak or ambiguous for them to come to that conclusion unaided.

This bias can be overcome by a sufficiently strong indication that there is an edge pointing in a different direction; but if the line gets faint, or part of it is missing, the bias automatically makes the brain respond as if the line was continuous. So the brain doesn't 'fill in' the gaps: it is set up not to notice that there are gaps. That's just one layer of the visual cortex, and it uses a rather simple trick: extrapolation. We have little idea, as yet, of the inspired guesswork that goes on in deeper layers of the brain, but we can be sure that it's even more clever, because it produces such a vivid sensation of a complete image.

What about hearing? How does that relate to sound? The standard lie-to-children about vision is that the cornea and lens make a picture on the retina, and that allegedly explains vision. Similarly, the corresponding lie-to-children about hearing centres on a part of the ear called the cochlea, whose structure allegedly explains how you analyse sound into different notes. In cross-section, the cochlea looks like a sliced snail-shell, and according to the lie-to-children, there are hair-cells all the way down the spiral attached to a tuned membrane. So different parts of the cochlea vibrate at different frequencies, and the brain detects which frequency – which musical note – it is receiving, by being told which part of the membrane is vibrating. In support of this explanation, we are told a rather nice story about boiler-makers, whose hearing was often damaged by the noise in the factories where they worked. Supposedly, they could hear all frequencies except ones near the frequency that was most common in making boilers. So just one place on their cochlea was burnt out, and the rest worked OK. This proved, of course, that the 'place' theory of hearing was correct.

Actually, this story tells you only how the ear can discriminate notes, *not* how you hear the noise. To explain that, it is usual to invoke the auditory nerve, which connects the cochlea to the brain. However, there are as many connections, or more, that go in the other direction, from brain to cochlea. *You have to tell your ear what to hear.*

Now that we can actually look at what the cochlea does when it's hearing, we find not one place vibrating for each frequency, but more like twenty. And these places move as you flex your outer ear. The cochlea is phase-sensitive, it can discriminate the kind of difference that makes an 'ooh' sound different from an 'eeh' at the same frequency. This is the kind of change to the sound that you make when you change the shape of your mouth as you speak. And surprise, surprise, that's just the difference that the cochlea – after your outer ear and your own particular auditory canal, and your own particular eardrum and those three little bones – can best discriminate. A recording from someone else's eardrum, played back up against yours, makes little sense. You have learned your own ears. But you have taught them, too.

There are about seventy basic sounds, called phonemes, that *Homo sapiens* uses in speech. Up to about six months old, all human babies can discriminate all of these, and an electrode on the auditory nerve gives different patterns of electrical activity for each. At about six to nine months old, we start talking scribble, and it very soon becomes English scribble or Japanese scribble. By a year old the Japanese ear cannot distinguish 'l' from 'r', because both phonemes send the same message from cochlea to brain. English babies can't discriminate the different clicks of the !Kung San, nor the differences between the distinct 'r's in French. So our sense organs do *not* show us the real world. They stimulate our brains to produce, to invent if you like, an internal world made of the counters, the Lego™ set, that each of us has built up as we mature.

Such apparently straightforward abilities as vision and hearing are far more complicated than we usually imagine. Our brains are much more than just passive recipients. An awful lot is going on inside our heads, and we project some of it back into what we think is the outside world. We are conscious only of a small part of its output. These hidden depths and strange associations in the brain may well be responsible for our musical sensibilities.

Music exercises the mind; it's a form of play. It seems probable that our liking for music is linked to other things than our ears. In particular, the brain's motor activity may be involved, as well as its sensory activity. In primitive tribes and advanced societies, music and dance often go together. So it may be the combination of sound and movement that appeals to our brains, rather than one or the other. In fact, music may be an almost accidental by-product of how our brains put the two together.

Patterns of movement have been common in our world for millions of years, and their evolutionary advantage is clear. The pattern 'climb a tree' can protect a savannah ape from a predator, and the same goes for the pattern 'run very fast'. Our bodies surround us with linked patterns of movement and sound. Like music, they are patterns in time, rhythms. Breathing, the heartbeat, voices in synch with lips, loud bangs in synch with things hitting other things.

There are common rhythms in the firing of nerve cells and the movement of muscles. Different gaits – the human walk and run, the walk-trot-canter-gallop of the horse – can be characterised by the timing with which different limbs move. These patterns relate to the mechanics of bone and muscle, and also to the electronics of the brain and the nervous system. So Nature has provided us with rhythm, one of the key elements of music, as a side-effect of animal physiology.

Another key element, pitch and harmony, is closely related to the physics and mathematics of sound. The ancient Pythagoreans discovered that when different notes sounded harmonious, there was a simple mathematical relationship between the lengths of the strings that produced them, which we now recognise as a relation between their frequencies. The octave, for example, corresponds to a doubling of frequency. Simple whole number ratios are harmonious, complicated relationships are not.

One explanation for this is purely physical. If notes with frequencies that are not related by simple whole numbers are sounded together, they interfere with each other to produce 'beats', a jarring low-frequency buzz. Sounds that make the sensory hairs in our ears vibrate in simple patterns are necessarily harmonious in the Pythagorean sense, and if they aren't, we hear the beats and they have an unpleasant effect. There are many mathematical patterns in musical scales, and they can be traced, to a great extent, to the physics of sound.

Overlaid on the physics, though, are cultural fashions and traditions. As a child's hearing develops, its brain fine-tunes its senses to respond to those sounds that have cultural value. This is why different cultures have different musical scales. Think of Indian or Chinese music compared to European; think of the changes in European music from Gregorian chants to Bach's *Well-Tempered Clavier*.

This is where the human mind is situated: on the one hand, subject to the laws of physics and the biological imperatives of evolution; on the other, as one small cog in the great machine of human society. Our liking for music has emerged from the interaction of these two influences. This is why music has clear elements of mathematical pattern, but is usually at its best when it throws the pattern book away and appeals to elements of human culture and emotion that are – for now, at least – beyond the understanding of science.

Let's come down to Earth and ask a simpler question. The wells of human creativity run deep, but if you take too much water from a well it runs dry. Once Beethoven had written the opening bars of his Symphony in C Minor – *dah-dah-da DUM* – that was one less tune for the rest of us. Given the amount of music that has been composed over the ages, maybe most of the best tunes have been found already. Will the composers of the future be unable to match those of the past because the world is running out of tunes?

There is, of course, far more to a piece of music than a mere tune. There is melody, rhythm, texture, harmony, development … But even Beethoven knew you can't beat a good tune to get your composition off the ground. By 'tune' we mean a relatively short section of music – what the *cognoscenti* call a 'motif' or a 'phrase', between one and thirty notes in length, say. Tunes are important, because they are the building blocks for everything else, be it Beethoven or Boyzone. A composer in a world that has run out of tunes is like an architect in a world that has run out of bricks.

Mathematically, a tune is a sequence of notes, and the set of all possible such sequences forms a phase space: a conceptual catalogue that contains not just all the tunes that have been written, but all the tunes that could ever be written. How big is T-space?

Naturally, the answer depends on just what we are willing to accept as a tune. It has been said that a monkey typing at random would eventually produce *Hamlet*, and that's true if you're willing to wait a lot longer than the total age of the universe. It's also true that along the way the monkey will have produced an incredible amount of airport novels.* In contrast, a monkey pounding the keys of a piano might actually hit on a reasonable tune every so often, so it looks as though the space of acceptably tuneful tunes is a reasonable-sized chunk of

---

* Though Ian has a friend, an engineer named Len Reynolds, whose cat managed to type 'FOR' into his computer by walking on the keyboard. Three more letters, 'MAT', and the cat would have wiped his hard disc.

the space of *all* tunes. And at that point, the mathematician's reflexes can kick in, and we can do some combinatorics again.

To keep things simple, we'll consider only European-style music based on the usual twelve-note scale. We'll ignore the quality of the notes; whether played on a piano, violin, or tubular bells, all that matters is their sequence. We'll ignore whether the note is played loudly or softly, and – more drastically – we'll ignore all issues of timing. Finally, we'll restrict the notes to two octaves, 25 notes altogether. Of course all these things are important in real music, but if we take them into account their effect is to *increase* the variety of possible tunes. Our answer will be an underestimate, and that's all to the good since it will still turn out to be *huge*. Really, *really* huge, right? No – bigger than *that*.

For our immediate purposes only, then, a tune is a sequence of 30 or fewer notes, each chosen from 25 possibilities. We can count how many tunes there are in the same way that we counted arrangements of cars and DNA bases. So the number of sequences of 30 notes is 25 × 25 × ... × 25, with 30 repetitions of that 25. Computer job, that: it says that the answer is

$$867361737988403547205962240695953369140625$$

which has 42 digits. Adding in the 29-note tunes, the 28-note ones, and so on we find that T-space contains roughly nine million billion billion billion billion tunes. Arthur C. Clarke once wrote a science fiction story about the 'Nine billion names of God'. T-space contains a million billion billion billion tunes for every one of God's names. Assume that a million composers write music for a thousand years, each producing a thousand tunes per year, more prolific even than The Beatles. Then the total number of tunes they will write is a mere trillion. This is such a tiny fraction of that 42-digit number that those composers will make no significant inroads into T-space at all. Nearly all of it will be unexplored territory.

Agreed, not all of the uncharted landscape of tune-space consists of *good* tunes. Among its landmarks are things like 29 repetitions of middle C followed by F sharp, and

BABABABABABABABABABABABABABABABABA,

which wouldn't win any prizes for musical composition. Nevertheless, there must be an awful lot of good new tunes still waiting to be invented. T-space is so vast that even if good-tune-space is only a small proportion of it, good-tune-space must also be vast. If all of humanity had been writing tunes non-stop since the dawn of creation, and went on doing that until the universe ended, we still wouldn't run out of tunes.

It is said that Johannes Brahms was walking along a beach with a friend, who was complaining that all of the good music had already been written. 'Oh, look,' said Brahms, pointing out to sea. 'Here comes the last wave.'

Now we come to what may well be the chief function of art and music for us – but not for edge people or chimpanzees, and probably not for Neanderthals. This, if we are right, is what Rincewind has in mind.

When we look at a scene we *see* only the middle five to ten degrees of arc. We invent the rest all around that bit, and we give ourselves the illusion that we're seeing about ninety degrees of arc. We perceive an extended version of the tiny region that our senses are detecting. Similarly, when we hear a noise, especially a verbal noise, we set it in a context. We rehearse what we've heard, we anticipate what's coming, and we 'make up' an extended present, as if we'd heard the whole sentence in one go. We can hold the entire sentence in our heads, as if we heard it *as* a sentence, and not one phoneme at a time.

This is why we can get the words of songs completely wrong and not realise it. The *Guardian* newspaper ran an amusing section on this habit, with examples such as 'kit-kat angel' for 'kick-ass angel' – bit of a generation gap there, which underlines how our perceptions are biased by our expectations. Ian recalls an Annie Lennox song that really went 'a garden overgrown with trees', but always sounded like 'I'm getting overgrown with fleas'.

Holding a whole sentence, or a musical phrase, in our minds is what we do with time when we watch a TV or a cinema-screen. We run the frames together into a series of scenes, as well as making up all the spatial stuff that we're not actually looking at. The brain has so many tricks that its owner is not conscious of: as you sit there in the cinema, your eyes are flicking from place to place on the screen, as they are doing while you read these words. But you turn off your perceptions as your eyes move, and re-jig your invented image so that your new retinal image is consistent with the previous version. That's why you get seasick or car-sick: if the outside image jumps about and isn't where you expect it to be, then that upsets your sense of balance.

Now think about a piece of music. Isn't the construction of an extended present precisely the exercise that your brain 'wants' to do with a series of sounds, but without the complication of the meanings? As soon as you get used to the style of a particular kind of music, you can listen to it and grasp whole themes, tunes, developments, even though you're hearing only one note at a time. And the instrumentalist who is making the noise is doing the same kind of thing. His brain has expectations of what the music should sound like, and he fulfils those. To some extent.

So it seems that our sense of music may be tied to a sense of an extended present. Some possible scientific evidence for this proposition has recently been found by Isabelle Peretz. In 1977 she identified a condition called 'congenital amusia'. This is not tone-deafness, but *tune*-deafness, and it should give us some insight into how normal people recognise tunes, by showing how that goes wrong. People with this condition cannot recognise tunes, not even 'Happy Birthday To You', and they have little or no sense of the difference between harmony and dissonance. There is nothing physically wrong with their hearing, however, and they *were* exposed to music as children. They are intelligent and have no history of mental illness. What seems to be wrong is that when it comes to music, they have no sense of an extended present. They cannot tap their feet in rhythm. They have no idea what a rhythm is. Their sense of timing is impaired. Mind you, so is their sense of pitch; they cannot distinguish sounds separated by an interval of two semitones – adjacent white keys on a piano. So the lack of an extended present is not the only problem. Congenital amusia is rare, and it affects males and females equally. Its sufferers have no difficulty with language, however, suggesting that the brain's music modules, or at least those affected by amusia, differ from its language modules.

The same kind of interpretational step takes place in the visual arts, too. When you look at a painting – a Turner, say – it evokes in you a variety of emotions, perhaps nostalgia for a nearly forgotten holiday on a farm. That may give you a little burst of endorphins, chemicals in the brain that create a sense of well-being, but presumably you'd get much the same from a photograph or even a verbal description or a bit of pastoral poetry. The Turner painting does more than that, perhaps because it can be more sentimental, more idealised than a photo, however idyllic. It evokes the memory on a more personal level.

What about other kinds of painting: the paper textures, the charcoal smear? Jack went to an art gallery, as an innocent in art appreciation, and tried the 'context' trick that any novice is always told to try. You're supposed to sit in front of the picture, and gaze at it, and kind of sink into it and feel how it relates to its surroundings. The result was instructive. When he paid attention to a small part of the canvas, he found that he could match the context that his brain had invented with the one that the artist had actually provided. The charcoal smear was particularly good for this: each part implied something of the pattern of the whole. However, there were intriguing differences from part to part. There were variations on the theme, as in music, superimposed on the brain's expectations. Jack's brain enjoyed comparing the picture that it was inventing with the progressively different one that the artist was forcing his brain to construct.

Art goes back a long, long way; the further back we look, the more controversial the evidence is. The 'Dame à la Capuche', a 1.5-inch (3.5-cm) high statuette of a woman, exquisitely carved from mammoth-tusk ivory, is 25,000 years old. Some of the most elegant cave paintings, with simple, sweeping lines that depict horses, bison and the like, are found in the Grotte Chauvet in France, and in 1995 they were dated at 32,000 years old. The oldest art that undoubtedly *is* art is about 38,000 years old: beads and pendants, found in Russia. And some beads made from ostrich egg shells in Kenya, which may be 40,000 years old.

Further back, it all gets less certain. Ochre is a common pigment in rock drawings, and ochre 'crayons' found in Australia are 60,000 years old. There is a lump of rock from the Golan Heights, whose natural crevices have been worn deeper, presumably by a human hand wielding another lump of rock. It bears a vague resemblance to a woman, and it is about 250,000 years old. But maybe it's just a lump of rock that a child idly scratched, and the shape is accidental.

Imagine yourself in the cave as the artist paints bison on the wall. He (or she?) is creating a picture for your brain that differs progressively from the one that your brain expects: 'Now let's put a female woolly rhinoceros under him …' There have been several 'artists' on television, doing precisely that trick. Rolf Harris was surprisingly good at drawing animal sketches before your very eyes. And they were iconic animals, too: sly fox and wise owl.

There it all is, tied up in a bundle. Our perceptions are tied to our expectations, and we do not segregate sensations from each other, or from memories. They are all played off against each other in the seclusion of our minds. We absolutely do not program our brains with direct representations of the real world. From the beginning we're instructing our brains what to make of what we see, hear, smell and touch. We put spin on everything, and we anticipate, compare and contrast, construct lengths of time from successive instants, construct areas of picture from focused observation. We've been doing this, layer upon layer, taking more subtle nuances from conversation, from flirting glances, to 'Will she come to look like her mother does now?' assessments of the real world, all the time.

That's what our brains do, and what edge people's brains don't.

We suspect that Neanderthals didn't do that kind of thing much, either, because there's an alternative, and it's consistent with their cultural torpor. The alternative is to live in a world that you've set up to ensure that nothing is unexpected. All the events follow your expectations from previous events, so habit engenders security. Such a world is very stable, and that means it doesn't *go* anywhere much. Why try to leave the Garden of Eden? Gorillas don't.

Tribal life could be like that for *Homo sapiens*, except that reality always intrudes, for instance those barbarians up on the hill. But

Neanderthals, maybe, weren't afflicted by barbarians. Certainly, nothing seems to have provoked big changes in their lifestyles, even over tens of thousands of years. Art does provoke changes. It makes us look at the world in new ways. The elves like that, it gives new ways for them to terrify people. But Rincewind has seen further than the elves are capable of seeing, and he's worked out where art takes us. Where? You'll soon find out.

TWENTY-FIVE
PARAGON OF VEGETABLES

 THE WINE-DARK SEA lapped the distant shore. Nice country, Rincewind thought. A bit like Ephebe. Grapes, olives, honey and fish and sunshine.

He turned back to his group of proto-actors. They were having difficulty grasping the idea.

'Like the priests do in the temples?' said a man. 'Is that what you mean?'

'Yes, but you can ... expand the idea,' said Rincewind. 'You can pretend to *be* the gods. Or anything else.'

'Wouldn't we get into trouble?'

'Not if you did it respectfully,' said Rincewind. 'And people would ... sort of *see* the gods. Seeing is believing, eh? Besides, children pretend to be other people all the time.'

'But that is childish play,' said the man.

'People might pay to see you,' said Rincewind. There was an immediate increase in interest. Human-shaped creatures were the same everywhere, Rincewind thought; if you got money for doing it, it had to be worth doing.

'Just gods?' said a man.

'Oh, no. Anything at all,' said Rincewind. 'Gods, demons, nymphs, shepherds—'

'No, I couldn't do a shepherd,' said the possible thespian. 'I'm a carpenter. I don't know shepherding.'

'But you know godding?'

'Well, yeah, that's just ... thundering and shouting and that kind of thing. Being a decent shepherd takes years of work.'

'You can't expect us to act like *people,*' said another man. 'That wouldn't be right.'

'It's not respectful,' said a third man.

Yes, we mustn't change things, thought Rincewind. The elves like that thinking. We mustn't change things, in case they end up different. Poor old Phocian …

'Well, can you do trees?' he said. He was vaguely aware that actors warmed up by pretending to be trees, amongst other things, and this presumably prevented wooden performances.

'Trees are all right,' said a man. 'They're quite magical. But it wouldn't be respectful to our friend over there to ask us to be carpenters.'

'All right, then, trees. That's a start. Now, stretch out your—'

There was a roll of thunder and a goddess appeared. Her hair was in golden ringlets, her white robe flapped in the breeze, and there was an owl on her shoulder. The men ran away.

'Well, my little trickster,' said the goddess, 'and what are you teaching them?'

Rincewind clapped a hand over one eye for a moment.

'That owl's *stuffed,*' he said. 'You can't fool me! No animals stay around elves without going mad!'

The image of the goddess wavered as the Queen tried to maintain control, but glamour is susceptible to disbelief.

'Oh, so brave?' she said, defaulting to her usual appearance.

She turned at a creaking noise behind her. The Luggage had tiptoed up and opened its lid.

'That doesn't frighten *me,*' she said.

'Really? It frightens me,' said Rincewind. 'Anyway, I'm simply brushing up their acting skills. Absolutely no problem there, is there? You should *love* these people. There's dryads, nymphs, satyrs, centaurs, harpies and big giants with one eye, unless that's a joke about sex I haven't fully understood yet. They believe in all of them and none of them exist! Except possibly the one-eyed giant, that one's a bit of a puzzler.'

'We have seen their performances,' said the Queen. 'They are not respectful of their gods.'

'But seeing is believing, isn't it? And you must admit, they've got a lot of gods. Dozens.'

He gave her a friendly smile, while hoping that she was keeping away from the local cities. They had a lot of temples in them, and shrines all over the place, but they also had a number of men who, while taking care to invoke the gods on every occasion, then appeared to expound ideas that didn't seem to have any place for gods in them, except as observers or decoration. But the actors liked playing gods …

'You're up to something,' said the Queen. 'Everywhere we look, you wizards are teaching people art. Why?'

'Well, it's a rather drab planet,' said Rincewind.

'Everywhere we go, they're telling stories,' said the Queen, still slowly circling. 'They're filling the sky with pictures, too.'

'Oh, the constellations?' said Rincewind. 'They don't change, you know. Not like at home. Amazing. I tried getting one tribe to name that big one – you know, with what looks like a belt? I thought if they ended up calling it the Bursar, and that group of little stars off to the right became The Dried Frog Pills, it'd be a nice souvenir of our visit—'

'You're frightened of me, aren't you,' said the Queen. 'All you wizards are frightened of women.'

'Not me!' said Rincewind. 'Women are less likely to be armed!'

'Yes you are,' the Queen insisted, moving closer. 'I wonder what your deepest desire is?'

Not to be here right now would be favourite, Rincewind thought.

'I wonder what I could give you,' said the Queen, caressing Rincewind's cheek.

'Everyone knows that anything you get from the elves is gone by morning,' said Rincewind, trembling.

'Yet many things are transient but pleasurable,' said the Queen, moving rather too close. 'What is it *you* want, Rincewind?'

Rincewind shuddered. There was no way he could lie.

'Potatoes,' he said.

'Tuberous vegetables?' said the Queen, her brows knitting in puzzlement.

'Well, yes. They've got them on one of the other continents, but they're not what *I'd* call spuds, and Ponder Stibbons says that if we left things as they were then by the time they've been brought over to this continent and bred up a bit it'd be the end of the world. So we thought we ought to ginger up the creativity level a bit.'

'And that's it? That's why all you wizards are doing all this? Just to accelerate the breeding of a *vegetable?*'

'*The* vegetable, thank you,' said Rincewind. 'And you did ask. The potato, in my opinion, is the crown of the vegetable kingdom. There's roast potatoes, jacket potatoes, boiled potatoes, fried potatoes, curried potatoes—'

'Just for a stupid tuber?'

'— potato soup, potato salad, potato pancakes —'

'All this for something that doesn't even see daylight!'

'—mashed potato, chipped potato, stuffed potato—'

The Queen slapped Rincewind's face. The Luggage bumped into the back of her legs. It wasn't entirely sure what was happening here. There were some things humans did that could be misinterpreted.

'Do you not think I could give you something better than a potato?' she demanded.

Rincewind looked puzzled.

'Are we talking about a sour cream topping with chives?' he said.

Something fell out of Rincewind's robe as he shifted uneasily. The Queen grabbed it.

'What's this?' she said. 'There's writing all over it!'

'It's just a script,' said Rincewind, still thinking about potatoes. 'A sort of story of a play,' he added. 'Nothing important at all. People going mad and getting killed, that sort of thing. And a glowworm.'

'I recognise this script! It's from the future of this world. Why would you carry it around? What is so special? Hah, are there potatoes in it?'

She leafed through the pages, as if she could read.

'This must be important!' she snapped. And vanished.

One solitary page slid down to the ground.

Rincewind bent down and picked it up. Then he shouted hotly at the empty air: 'I suppose a packet of crisps is out of the question?'

TWENTY-SIX
LIES TO CHIMPANZEES

 A CENTRAL FEATURE OF HUMAN INTELLIGENCE is the ability to infer what is going on in another person's mind, to guess what the world looks like from *their* point of view. Which is what Rincewind is trying to stop the Queen of the Elves from doing. We can't make such inferences with perfect accuracy; that would be telepathy, which is almost certainly impossible, because each brain is wired up differently and therefore represents the universe in its own special way. But we've evolved to be pretty good at guessing.

This ability to get inside other people's heads has many beneficial consequences. One is that we recognise other people as people, not just automata. We recognise that they *have* a mind, that to them the universe seems just as real and vivid as it does to us, but that the vivid things they perceive may not be the same as those that we perceive. If intelligent beings are going to get along together without too much friction, it's important to realise that other members of your species have an internal mental universe, which controls their actions in the same way that your own mind controls yours.

When you can put yourself inside another person's mind, stories gain a new dimension. You can identify with a central character, and vicariously experience a different world. This is the appeal of fiction: you can captain a submarine, or spy on the enemy, from the safety and comfort of an armchair.

Drama has the same appeal, too, but now there are real people to identify with; people who play a fictional role. Actors, actresses. And they rely even more on getting inside other people'ss minds, especially the minds of fictional characters. Macbeth. The Second Witch. Oberon. Titania. Bottom.

How did this ability arise? As usual, it seems to have come about because of a complicity between the internal signal-processing abilities of the brain and the external pressures of culture. It arose through an evolutionary arms race, and the main weapon in that race was the lie.

The story starts with the development of language. As the brains of proto-humans evolved, getting larger, there was room in them for more kinds of processing tasks to be carried out. Primitive grunts and gestures began to be organised into a relatively systematic code, able to represent aspects of the outside world that were important to the creatures concerned. A complicated concept like 'dog' became associated with a particular sound. Thanks to an agreed cultural convention, anyone who heard that sound responded to it with the mental image of a dog; it wasn't just a funny noise. If you try to listen to someone speaking a language that you know, focusing just on the noises that they are making and trying not to pick up the meaning of their words, you'll find that it's almost impossible. If they speak a language far removed from any that you know, however, their speech comes over as a meaningless gabble. It conveys *less* to you than a cat's miaow.

In the brain are circuits of nerve cells that have learned to decode gabble into meaning. We've seen that as a child grows, it begins by babbling a random assortment of phonemes, the 'units' of sound that a human mouth and larynx can produce. Gradually the child's brain prunes the list down to those sounds that it hears from its parents and other adults. While it is doing that, the brain is *destroying* connections between nerve cells that seem to be obsolete. Quite a lot of the early mental development of an infant consists of chopping down a randomly connected, all-purpose brain, and pruning it into a brain that can detect the things that are considered important in the child's culture. If the child is not exposed to much linguistic stimulus in early childhood – such as a 'feral' child brought up by animals – then they can't learn a language properly in later life. After about the age of ten, the brain's ability to learn language fades away.

Much the same happens with other senses, in particular the sense of smell. Different people smell the same thing differently. To some, a particular odour may be offensive, to others innocuous, and to yet others, nonexistent. As with language, there are cultural biases to certain smells.

The primary function of language – by which we mean 'the main evolutionary trick that made it advantageous, leading to its preservation and enhancement by natural selection' – is to convey meaningful messages to other members of the same species. We do this in several ways: 'body language' and even bodily odours convey vivid messages, largely without our being conscious of them. But spoken language is far more versatile and adaptable than the other kinds, and we are very conscious of what others are saying. Especially when it is about *us*.

One of the commonest generic evolutionary tricks is to cheat. As soon as a bunch of organisms has evolved some specific ability or behaviour, a new possibility arises: subverting that behaviour. Predictable behaviour patterns provide a natural springboard from which organisms can leap out into the space of the adjacent possible. Bees evolved the abilities to collect nectar and pollen, to feed themselves. Later, we subverted that activity by providing them with better homes than they would find in nature. We get to steal their honey, by providing them with hives as the up-market adjacent-possible homes.

Many evolutionary trends have arisen from subversion. So, as the ability to put specific thoughts into the minds of others became established, it was natural for evolution to experiment with methods for subverting that process. You didn't have to put your own genuine thoughts into the minds of others: you could try to put different thoughts there. Perhaps you could gain an advantage by misleading the creatures you were 'communicating' with. The result was the evolution of lying.

Many animals tell lies. Monkeys have been observed making the troupe's 'danger' call-sign. Then, as the rest of the troupe heads off for cover, the liar grabs the food that they have temporarily abandoned. On a more primitive but just as effective level, mimicry in the animal kingdom is a form of lying. A harmless hover-fly displays the black-and-yellow warning bands of a wasp, telling the lie 'I am dangerous, I can sting'.

As humanity evolved, those monkey lies turned into more sophisticated ape lies, then hominid lies, then human lies. As we became more intelligent, our capacity for telling lies co-evolved alongside another important ability: the ability to tell when someone was lying to you. A monkey troupe can evolve several defences against a member who abuses the danger-signal for his own ends. One is to recognise that this individual can't be trusted, and ignore their calls. The nursery tale of the little boy who cried 'wolf' exposes the dangers inherent in this area, both for the troupe and for the individual. Another is to punish the individual for telling the lie. A third is to evolve the ability to tell the difference between a lying danger-signal and a true one. Is the monkey crying 'danger' staring at someone else's food with a greedy glint in their eye?

Just as there are sound evolutionary reasons for telling lies, so there are sound evolutionary reasons for being able to detect them. If others are trying to manipulate you to their advantage, then it is very probably to your disadvantage. So it is in your best interests to realise that, and avoid being manipulated. The result is an inevitable arms race, in which the ability to tell lies is played off against the ability to detect them. It is no doubt still going on, but already the result is some very sophisticated lying, and some very sophisticated detection. Sometimes the look on a person's face tells us they're telling an untruth; sometimes the tone of voice.

One effective way to recognise a lie is to put yourself inside the other person's mind, and ask yourself whether what they are saying is consistent with what you have convinced yourself they are thinking. For instance, they are saying what a sweet little child you have, but you remember from previous encounters that usually they can't stand kids. Maybe your child is different, of course, but then you notice that worried look in their eyes, as if they'd rather be somewhere else …

Empathy is not just a nice way to understand someone else's point of view. It's a weapon that you can use to your own advantage. Having understood their point of view, you can compare it with what they're saying, and work out whether to believe them. In this manner, the existence of lies in language's phase space of the adjacent possible encouraged the development of human empathy, and with it, individual intelligence and collective social cohesion. Learning to tell lies was a major step forward for humanity.

We can put ourselves inside the minds of other people with some degree of credibility, because we are people ourselves. We do at least know what it's like to be a person. But even then, we are probably deluding ourselves if we think that we really know exactly what's going on inside someone else's mind, let alone what that feels like to them. Each human mind is wired differently, and is the product of its owner's own experiences. It is even more problematic whether we can imagine what it is like to be an animal. On Discworld, an accomplished witch can put herself inside an animal's mind, as we see, for instance, in this passage from *Lords and Ladies*:

> She Borrowed. You had to be careful. It was like a drug. You could ride the minds of animals and birds, but never bees, steering them gently, seeing through their eyes. Granny Weatherwax had many times flicked through the channels of consciousness around her. It was, to her, part of the heart of witchcraft. To see through other eyes ...
>
> ... through the eyes of gnats, seeing the slow patterns of time in the fast pattern of one day, their minds travelling rapidly as lightning ...
>
> ... to listen with the body of a beetle, so that the world is a three-dimensional pattern of vibrations ...
>
> ... to see with the nose of a dog, all smells now colours ...

It's a poetic image. Does a dog 'see' smells? There is a folk belief that smell is far more important to a dog than sight, but this could well be an exaggeration based on the more credible observation that smell is more important to dogs than it is to humans. But even here we must add 'consciously, at least', because we react subconsciously to pheromones and other emotionally loaded chemicals. Some years ago David Berliner was working on the chemicals in human skin, and he left an open beaker containing some skin extracts on the laboratory bench. Then he noticed that his lab assistants were becoming distinctly more animated than usual, with a lot of camaraderie and mild flirtation. He froze the extract and put it away in the laboratory refrigerator for safekeeping. Thirty years later, he analysed the substances in the beaker and found a chemical called androstenone, which is rather like a sex hormone. A series of experiments showed that this chemical was responsible for the animated behaviour. However, androstenone has no smell. What was going on?

Some animals possess a 'vomeronasal' organ (often called the 'second nose'). This is a small region of tissue in the nose, which detects certain chemicals but is separate from the standard olfactory (smelling) system. The conventional wisdom had long been that humans do not possess a vomeronasal organ, but the curious behaviour of his assistants made the scientist wonder. Berliner discovered that the conventional wisdom was wrong: some humans, at least, do have a vomeronasal organ, and it responds to pheromones. Those are special chemicals that trigger strong responses in animals, such as fear or sexual arousal. The vomeronasal organ's owners are not consciously aware that they are sensing anything, but boy, do they respond.

This story shows how easily we can get sensations wrong. In this case, you *know* what it vomeronasally smells like to be a human: you don't feel anything at all, not consciously. But you certainly respond! So your reactions, and what they 'feel like', are very different. The sounds we hear, the sensations of heat and cold on our skin, the smells that assail our nostrils, the unmistakable taste of salt ... all these are qualia, vivid 'feelings' stuck on to our perceptions by our minds to help us recognise them more readily. They have a basis in reality, yes, but they are not real features of the outside world. They must be real features of brain architecture and function, real things happening in real nerve cells, but that level of reality is very different from the level that we perceive.

So we should be suspicious of the belief that we can know what it feels like to be a dog. In 1974 the philosopher Thomas Nagel published a famous essay 'What is it like to be a bat?' in the *Philosophical Review*, in which he made the same point. We can imagine what it is like to be a human who is behaving – superficially at least – like a bat, but we have no idea what it feels like to the *bat*, and it is questionable whether human knowledge can ever extend in such a direction.

We probably get bats wrong anyway. We know that bats use echo-location to sense their surroundings, much as a submarine uses sonar. The bat or submarine emits sharp pulses of sound, and hears the returning echoes. From those, it can 'compute' what the sound must be bouncing off. We naturally assume that the bat responds to echoes in the same kind of way that we would: it hears them. We naturally expect the qualia of bat echo-location to be similar to the human qualia evoked by sound-patterns, of which the richest example is music. So we imagine the bat flying along to the accompaniment of incredibly rapid rhythms played on bongo drums.

However, this could be a false analogy. Echo-location is the main sense of a bat, so the 'correct' corresponding sense of a human is *its* main sense, which is sight, not hearing. The August 1993 edition of *Nature* has a picture of a bat on the cover, with the words 'How bats' ears see'. This refers to a technical article, by Steven Dear, James Simmons and Jonathan Fritz, who discovered that the neurons in the part of the bat's brain that processes returning echoes are connected together in a very similar way to those in the human *visual* cortex. In terms of neural architecture, it looks very much as if the bat's brain uses the echoes to build up an *image* of its surroundings. Analogously, today's submarines use computers to turn a series of echoes into a three-dimensional map of the surrounding water. *Figments of Reality* developed this point to give a partial answer to Nagel's question:

[In effect] bats *see* with their ears, and their sonar qualia might well be like our visual ones. Intensity of sound might come over to the bat as a kind of 'brightness', and so on. Possibly the bat's sonar qualia 'see' the world in black and white and shades of grey, but they could also pick up and render vivid various more subtle features of sound reflections. The closest analogy in humans is texture, which we sense by touch, but the bat could sense by sound. Soft objects reflect sound less well than hard ones, for instance. So bats may well 'see' textured sound. If so – and here our analogy is intended only as a very rough way to convey the idea – the sonar quale for a soft surface might 'look' green to the bat's mind, that for hard ones might look red, that for liquid ones like a colour only bees can see, and so on …

On Roundworld, such statements are no more than guesswork, supported by analogies of neural architecture. On Discworld, witches know what it feels like to be a bat, or a dog, or a beetle. And Angua the werewolf smells in colours, which is very close to our suggestion that bats hear in images and 'see' textures. But even on Discworld, the witches do not actually feel what it is like to be a bat. They feel what it is like to be a human who has 'borrowed' the sensory organs and neural processing equipment of a bat. It may feel quite different to be a bat when a witch is not hitching a free ride on its mind.

Even though we can't be certain what it feels like to be an animal, or another person, the attempt has several uses. As we said, the ability involved here is empathy: being able to understand what another person feels like. We've already seen that this is an important social skill, and that the same ability, deployed in a different way for a different purpose, gives us a chance to detect that someone else is lying to us. If we put ourselves inside their heads and realise that what they are saying is different from what we believe they are thinking, then we suspect them of lying.

The word 'lie' has negative overtones, deservedly so, but what we're talking about here can be constructive as well as destructive, and often is. For the purposes of the present discussion, a lie is anything contrary to the truth, but it's not at all clear what 'the truth' is, or even whether there is only one of it, as the word 'the' would seem to indicate. When two people have a row, it is generally impossible for either of them, or anyone else, to figure out exactly what really happened. Our thoughts are tainted by perceptions. This is unavoidable, because what we think of as being 'real' is what our minds make of what comes from the sense organs: fudged, tuned, and mangled by a succession of interpretations by different bits of brain, *plus* some wallpaper additions. We never know what is *really* out there around us. All we know is what our minds construct from what our eyes, ears and fingers report.

Not to put too fine a point on it, those perceptions are lies. The vivid universe of colour that our brain derives from the light that falls on our retinas *does not really exist*. The redness of a rose is derived from its physical features, but 'being red' is not a physical feature as such. 'Emitting light of a certain wavelength' gets closer to being a physical feature. However, the vivid redness that we 'see' does not correspond to a specific wavelength. Our brains correct the colours of visual images for shadows, light reflected on to parts of the image by other parts of a different colour, and so on. Our sensation of redness is a decoration added to the perception by our brain: a quale. So what we 'see' is not an accurate perception of what is there, but a mental transform of a sensory perception of what is there.

To a bee, that same uniformly red rose may look very different, with obvious markings. The bee 'sees' in ultraviolet, a wavelength outside our range of perceptions. The rose emits a whole distribution of wavelengths of light; we see a small part of that, and call it reality. The bee sees a different part and responds to it in its own beelike way, using the markings to land on the flower and collect nectar, or to dismiss it from consideration and fly on to the next possibility. Neither the bee's perception, nor ours, *is* the reality.

In Chapter 24 we explained that our minds select what they perceive in more ways than just passively ignoring signals that our senses can't pick up. We fine-tune our senses to see what we want them to see, hear what we want them to hear. There are more nerve connections going *from* the brain to the ear than there are from the ear to the brain. Those connections adapt the ear's ability to perceive certain sounds, maybe by making it more sensitive to sounds that could represent danger and less sensitive to sounds that don't really matter much. People who are not exposed to certain sounds as children, when their ears and brains are being tuned to pick up language, cannot distinguish them as adults. To the Japanese, the two phonemes 'l' and 'r' sound identical.

The lies that our senses tell us are not malicious. They are partial truths rather than untruths, and the universe is so complicated, and our minds are so simple in comparison, that the best we can ever hope for is half-truths. Even the most esoteric 'fundamental' physics is at best a half-truth. Indeed, the more 'fundamental' it becomes, the less true it gets. It is therefore no surprise that the most effective method we have yet devised for passing extelligence on to our children is a systematic series of lies.

It is called 'education'.

We can hear the hackles rising even as we write, as quantum signals echo back down the timelines from future readers in the teaching profession turning to this page. But before hurling the book across the room or sending an offended e-mail to the publisher, ask yourself just how much of what you tell children is true. Not worthy, not defensible: *true*. At once you'll find yourself on the defensive: 'Ah, yes, but of course children can't understand all of the complexities of the real world. The teacher's job is to simplify everything as an aid to understanding ...'

Quite so.

Those simplifications are lies, within the meaning we are currently attaching to that word. But they are helpful lies, constructive lies, lies that even when they are really *very* wrong still open the door to a better understanding next time round. Consider, for example, the sentence 'A hospital is a place where people are sent so that the doctors can make them better'. Well, no sensitive adult would wish to tell a child that sometimes people go into hospital alive and come out dead. Or that often it's not possible to make them better. For a start, the child may have to go into hospital at some stage, and too big a dose of truth early on might make it difficult for the parents to persuade them to do so without making a fuss. Nonetheless, no adult would consider that sentence to be an accurate statement of what hospitals are really about. It is, at best, an ideal to which hospitals aspire. And when we justify our description on the grounds that the truth would upset the child, we are admitting that the sentence is a lie, and asserting that social conventions and human comfort are more important than giving an accurate description of what the world is about.

They often are, of course. A lot depends on context and intention. In Chapter 4 of *The Science of Discworld* we called these helpful untruths and half-truths 'lies-to-children'. They must be distinguished from the much less benevolent 'lies-to-adults', another word for which is 'politics'. Lies-to-adults are constructed with the express purpose of concealing intentions; their aim is to mislead. Some newspapers tell lies-to-adults; others do their best to tell truths-to-adults, although they always end up by telling adult versions of lies-to-children.

In the twenty-fifth Discworld novel *The Truth*, journalism comes to the Disc, in the form of William de Worde. His career begins with a monthly newsletter sent to various Discworld notables, usually for five dollars each month, but in the case of one foreigner for half a cartload of figs twice per year. He writes one letter, and pays Mr Cripslock the engraver in the Street of Cunning Artificers to turn it into a woodcut, from which he prints five copies. From these small beginnings emerges Ankh-Morpork's first newspaper, when de Worde's ability to sniff out a story is allied to the dwarves' discovery of movable type. It is rumoured that the dwarves have found a way to turn lead into gold – and since the type is made of lead, in a way they have.

The main journalistic content of the novel is a circulation battle between de Worde's *Ankh-Morpork Times*, with its banner 'THE TRUTH SHALL MAKE YOU FREE', and the *Ankh-Morpork Inquirer* (THE NEWS YOU ONLY HEAR ABOUT). The *Times* is an upmarket broadsheet, running stories with headlines like 'Patrician Attacks Clerk With Knife (He had the knife, not the clerk)', and checking its facts before publishing them. The *Inquirer* is a tabloid, whose headlines are more of the 'ELVES STOLE MY HUSBAND' kind, and it saves money by making all the stories up. As a result, it can undercut its upmarket competitor when it comes to price, *and* the stories are much more interesting. Truth eventually prevails over cheap nonsense, however, and de Worde learns from his editor Sacharissa a fundamental principle of journalism:

'Look at it like this,' said Sacharissa, starting a fresh page. 'Some people are heroes. And some people jot down notes.'

'Yes, but that's not very—'

Sacharissa glanced up and flashed him a smile. 'Sometimes they're the same person,' she said.

This time it was William who looked down, modestly.

'You think that's really true?' he said.

She shrugged. 'Really true? Who knows? This is a newspaper, isn't it? It just has to be true until tomorrow.'

Lies-to-children, even the broadsheet newspaper sort, are mostly benign and helpful, and even when they are not, they are *intended* to be that way. They are constructed with the aim of opening a pathway that will eventually lead to more sophisticated lies-to-children, reflecting more of the complexities of reality. We teach science and art and history and economics by a series of carefully constructed lies. Stories, if you wish ... but then, we've already characterised a story as a lie.

The science teacher explains the colours of the rainbow in terms of refraction, but slides over the *shape* of the rainbow and the way those colours are arranged. Which, when you come to think of it, are more puzzling, and more what we want to know about when we ask why rainbows look like they do. There's a lot more to the physics than a raindrop acting as a prism. Later, we may develop the next level of lie by showing the child the elegant geometry of light rays as they pass through a spherical raindrop, refracting, reflecting, and refracting back out again, with each colour of light focused along a slightly different angle. Later still, we explain that light does not consist of rays at all, but electromagnetic waves. By university, we are telling undergraduates that those waves aren't really waves at all, but tiny quantum wave-packets, photons. Except that the 'wave-packets' in the textbooks don't actually do the job ... And so on. *All* of our understanding of nature is like this; none of it is Ultimate Reality.

TWENTY-SEVEN
LACK OF WILL

THE WIZARDS WERE NEVER QUITE CERTAIN where they were. It wasn't their history. History gets named afterwards: The Age of Enlightenment, the Depression. Which is not to say that people sometimes aren't depressed with all the enlightenment around them, or strangely elevated during otherwise grey times. Or periods are named after kings, as if the country was *defined* by whichever stony-faced cut-throat had schemed and knifed his way to the top, and as if people would say, 'Hooray, the reign of the House of Chichester – a time of deep division along religious lines and continuing conflict with Belgium – is now at an end and we can look forward to the time of the House of Luton, a period of expansion and the growth of learning! The ploughing of the big field is going to be a lot more interesting from now on!'

The wizards had settled for calling the time they'd arrived 'D' and, now, they were back there, in some cases quite suntanned.

They had commandeered Dee's library again.

'Stage One seemed to have worked quite well, gentlemen,' said Ponder Stibbons. 'The world is certainly a lot more colourful. We do seem to have, er, *assisted* the elves in the evolution of what I might venture to call *Homo narrans*, or "Storytelling Man".'

'There's still religious wars,' said the Dean. 'And still the heads on spikes.'

'Yes, but for more interesting reasons,' said Ponder. 'That's humans for you, sir. Imagination is imagination. It gets used for *everything*. Wonderful art and really dreadful instruments of torture. What was that country where the Lecturer in Recent Runes got food poisoning?'

'Italy, I think,' said Rincewind. 'The rest of us had the pasta.'

'Well, it's full of churches and wars and horrors and some of the most amazing art. Better than we've got at home. We can be proud of that, gentlemen.'

'But when we showed them the book the Librarian found in L-space, of Great Works of Art with the full colour pictures ...' mumbled the Chair of Indefinite Studies, as if he had something on his mind but wasn't certain how to phrase it.

'Yes?' said Ridcully.

'... well, it wasn't actually *cheating*, was it?'

'Of course not,' said Ridcully. 'They must have painted them *somewhere*. Some other dimension. Something quantum. A parallel eventuality or something with that sort of a name. But that doesn't matter. It all goes round and round and it comes out here.'

'But I think we said too much to that big chap with the bald head,' said the Dean. 'The artist, remember? Could've been the double of Leonard of Quirm? Beard, good singing voice? You shouldn't have told him about the flying machine that Leonard built.'

'Oh, he was scribbling so much stuff no one'll take any notice,' said Ridcully. 'Anyway, who'll remember an artist who can't get a simple smile right? The point *is*, gentlemen, that the fantastic imagination and the, er, practical imagination go hand in hand. One leads to the other. Can't separate them with a big lever. Before you can make something, you have to picture it in your head.'

'But the elves are still here,' said the Lecture in Recent Runes. 'All we've done is do their work even better! I don't see the *point*!'

'Ah, that's Stage Two,' said Ponder. 'Rincewind?'

'What?'

'You're going to talk about Stage Two. Remember? You told us you wanted to get the world to the right stage?'

'I didn't know I had to make a presentation!'

'You mean you don't have any slides? No paperwork at all?'

'Paperwork slows me down,' said Rincewind. 'But it's obvious, isn't it? We say Seeing is Believing ... and I thought about that, and it's not really true. We don't believe in chairs. Chairs are just things that exist.'

'So?' said Ridcully.

'We don't *believe* in things we can see. We believe in things that we can't see.'

'And?'

'And I've been checking this world against L-space and I think we've made it the one where humans survive,' said Rincewind. 'Because now they can picture gods and monsters. And when you can picture them, you don't need to *believe* in them any more.'

After a long silence the Chair of Indefinite Studies said, 'Is it just me, or has anyone else noticed how many huge cathedrals they've been building on this continent? Big, big buildings full of wonderful crafts-manship? And those painters we talked to have been very keen on religious paintings …'

'And your point is … ?' said Ridcully.

'It's just that this has been happening at the same time as people have been really taking an interest in how the world works. They're asking more questions. How? and Why? and questions like that,' said the Chair of Indefinite Studies. 'They're acting like Phocian but with-out going mad. Rincewind seems to be suggesting that we're killing off the gods of this place.'

The wizards looked at him.

'Er,' he went on, 'if you think a god is huge and powerful and every-where, then it's natural to be god-fearing. But if someone comes along and paints that god as a big bearded chap in the sky, it's not going to be long before people say, don't be silly, there can't be a big bearded man on a cloud somewhere, let's go and invent Logic.'

'Can't there be gods here?' said the Lecturer in Recent Runes. 'We've got a mountaintop *full* of 'em at home.'

'We've never detected deitygen in this universe,' said Ponder thoughtfully.

'But it's said to be generated by intelligent creatures, just like cows generate marsh gas,' said Ridcully.

'In a universe based on magic, certainly,' said Ponder. 'This one is just based on bent space.'

'Well, there's been lots of wars, lots of deaths and I'd bet there's lots of believers,' said the Chair of Indefinite Studies, now looking extremely uncomfortable. 'When thousands die for a god, you get a god. If someone is *prepared* to die for a god, you get a god.'

'At home, yes. But does that work here?' said Ponder.

The wizards sat in silence for a while.

'Are we going to get into any sort of religious trouble for this?' said the Dean.

'None of us has been struck by lightning yet,' said Ridcully.

'True, true. I just wish there was a less, er, permanent test,' said the Chair of Indefinite Studies. 'Er ... the dominant religion on this continent seems to be a family concern, somewhat similar to Old Omnianism.'

'Big on smiting?'

'Not lately. It's gone very quiet *vis-à-vis* heavenly fire, widespread flooding and transmutation into food additives,' said the Chair.

'Don't tell me,' said Ridcully. 'A public appearance, some simple moral precepts, and then apparent silence? Apart, that is, for millions of people arguing what "Do not steal" and "Don't Commit Murder" actually mean?'

'That's right.'

'*Just* like Omnianism, then,' said the Archchancellor glumly. 'Noisy religion, silent god. We must tread carefully, gentlemen.'

'But I did point out that there is no perceptible trace whatsoever of any deities of any kind anywhere in this universe!' said Ponder.

'Yes, very puzzling,' said Ridcully. 'Nevertheless, we have no magical powers here and it pays to be careful.'

Ponder opened his mouth. He wanted to say: We know everything about this place! We've watched it happen! It's all balls, spinning in curves. It's matter bending space and space moving matter. Everything here is the result of a few simple rules! That's all! It's all just a matter of rules! It's all ... logical.

He wanted it to be logical. Discworld wasn't logical. Some things happened on the whim of gods, some things happened because it was a good idea at the time, some things happened out of sheer randomness. But there was no logic – at least, no logic that Ponder approved of. He'd gone to the little town called Athens that Rincewind had talked about, in a sheet borrowed from Doctor Dee, and listened to men not entirely unlike the philosophers of Ephebe talking about logic, and it had made him want to burst into tears. *They* didn't have to live in a place where things changed on a whim.

Everything ticked and tocked and turned for them like a great big machine. There were rules. Things stayed the same. The same reliable stars came up every night. Planets didn't disappear because they've wandered too close to a flipper and been flicked far away from the sun.

No trouble, no complications. A few simple rules, a handful of elements … it was all so easy. Admittedly, he found it a little hard to work out *exactly* how you got from a few simple rules to, say, the sheen on mother-of-pearl or the common porcupine, but he was sure that you did. He wanted, intensely, to believe in a world where logic worked. It was a matter of faith.

He envied those philosophers. They nodded to their gods and then, by degrees, destroyed them.

And now he sighed.

'We've done the best we can,' he said. 'Your plan, Rincewind?'

Rincewind stared at the glass sphere that was the current abode of Hex.

'Hex, is this world ready for the William Shakespeare of whom we spoke?'

'It is.'

'And he exists?'

'No. Two of his grandparents did not meet. His mother was never born.'

In his hollow voice, Hex recounted the sad history, in detail. The wizards took notes.

'Right,' said Ridcully, rubbing his hands together when Hex finished. 'This at least is a simple problem. We shall need a length of string, a leather ball of some kind, and a large bunch of flowers …'

Later, Rincewind stared at the glass sphere that was the current abode of Hex.

'Hex, now is this world ready for the William Shakespeare of whom we spoke?'

'It is.'

'And he exists?'

'Violet Shakespeare exists. She married Josiah Slink at the age of six-teen. No plays have been written, but there have been eight children of which five have survived. Her time is fully occupied.'

The wizards exchanged glances.

'Perhaps if we offered to babysit?' said Rincewind.

'Too many problems,' said Ridcully firmly. 'Still it's a change to have an *easy* one for once. We will need the probable date of conception, a stepladder and a gallon of black paint.'

Rincewind stared at the glass sphere that was the current abode of Hex.

'Hex, is this world ready for the William Shakespeare of whom we spoke?'

'It is.'

'And he exists?'

'He was born, but died at the age of 18 months. Details follow ...'

The wizards listened. Ridcully looked thoughtful for a moment.

'This will require some strong disinfectant,' he said. 'And a lot of carbolic soap.'

Rincewind stared at the glass sphere that was the current abode of Hex.

'Hex, is this world ready for the William Shakespeare of whom we spoke?'

'It is.'

'And he exists?'

'No. He was born, successfully survived several childhood illnesses, but was shot dead one night while poaching game at the age of thirteen. Details follow ...'

'Another easy one,' said Ridcully, standing up. 'We shall need ... let me see ... some drab clothing, a dark lantern and a very large cosh ...'

Rincewind stared at the glass sphere that was the current abode of Hex.

'Hex, is this world ready for the William Shakespeare of whom we spoke? *Please?*'

'It is.'

'And he exists?'

'Yes.'

The wizards tried not to look hopeful. There had been too many false dawns in the last week.

'Alive?' said Rincewind. 'Male? Sane? Not in the Americas? Not struck by a meteorite? Not left incapacitated by a hake during an unusual fall of fish? Or killed in a duel?'

'No. At this moment he is in the tavern that you gentlemen frequent.'

'Does he have all his arms and legs?'

'Yes,' said Hex. 'And … Rincewind?'

'Yes?'

'As one of two unexpected collateral events to this latest interference, the potato has been brought to these shores.'

'Hot damn!'

'And Arthur J. Nightingale is a ploughman and never learned to write.'

'Near miss there,' said Ridcully.

TWENTY-EIGHT
WORLDS OF IF

 THE WIZARDS HAVE DEVISED A SECRET WEAPON in their battle against the elves for the soul of Roundworld, and they are busily re-engineering history to make sure that their weapon gets invented. The weapon is one Will Shakespeare – Arthur J. Nightingale just can't hack it. And they're proceeding by trial and error, with a lot of both. Nonetheless, they gradually persuade the flow of history to converge, step by step, towards their desired outcome.

Black paint? You may know this superstitious practice, but if not: painting the kitchen ceiling black is supposed to guarantee a boy.* The wizards will try anything. To begin with. And if it doesn't work, they'll try something else, until eventually they get somewhere.

Why is it unreasonable to expect them to succeed in one go, but reasonable to expect them to achieve their objective by repeated refinements?

History is like that.

There is a dynamic to history, but we find out what that dynamic is only as the events concerned unfold. That's why we can put a name to historical periods only after they've happened. That's why the history monks on Discworld have to wander the Disc making sure that historical events that ought to happen *do* happen. They are the

---

* The superstition is common in the Black Country, in places like Wombourne and Wednesbury. Though that's not why it's called the Black Country. The thing about your Black Country is, it's black. At least, it *was* black, with industrial grime and pollution, when it got its name. Some bits no doubt still are.

guardians of narrativium and they spread it around dispassionately to ensure that the whole world obeys its storyline. The history monks come into their own in *Thief of Time*. Using great spinning cylinders called Procrastinators, they borrow time from where it is not needed and repay it where it is:

> *According to the Second Scroll of Wen the Eternally Surprised*, Wen the Eternally Surprised sawed the first procrastinator from a trunk of a *wamwam* tree, carved certain symbols on it, fitted it with a bronze spindle, and summoned the apprentice, Clodpool.
>
> 'Ah, very nice, master,' said Clodpool. 'A prayer wheel, yes?'
>
> 'No, this is nothing like as complex,' said Wen. 'It merely stores and moves time.'
>
> 'That simple, eh?'
>
> 'And now I shall test it,' said Wen. He gave it a half-turn with his hand.
>
> 'Ah, very nice, master,' said Clodpool. 'A prayer wheel, yes?'
>
> 'No, this is nothing like as complex,' said Wen. 'It merely stores and moves time.'
>
> 'That simple, eh?'
>
> 'And now I shall test it,' said Wen. He moved it a little less this time.
>
> 'That simple, eh?'
>
> 'And now I shall test it,' said Wen. This time he twisted it gently to and fro.
>
> 'That si-si-si That simple-ple, ëh eheh simple, eh?' said Clodpool.
>
> 'And I have tested it,' said Wen.

On Roundworld we don't have history monks – or, at least, we've never caught anyone playing that role, but could we ever do so? – but we do have a kind of historical narrativium. We have a saying that 'history repeats itself' – the first time as comedy, the second time as tragedy, because the one thing we learn from history is that we never learn from history.

Roundworld history is like biological evolution: it obeys rules, but even so, it seems to make itself up as it goes along. In fact, it seems to make up its *rules* as it goes along. At first sight, that seems incompatible with the existence of a dynamic, because a dynamic is a rule that takes the system from its present state to the next one, a tiny instant into the future. Nonetheless, there must be a dynamic, otherwise historians would not be able to make sense of history, even after the event. Ditto evolutionary biology.

The solution to this conundrum lies in the strange nature of the historical dynamic. It is emergent. Emergence is one of the most important, but also the most puzzling, features of complex systems. And it is important for this book, because it is the existence of emergent dynamics that leads humans to tell stories. Briefly: if the dynamic wasn't emergent, then we wouldn't need to tell stories about the system, because we'd all be able to understand the system on its own terms. But when the dynamic is emergent, a simplified but evocative story is the best description that we can hope to find …

But now we're getting ahead of our own story, so let's back up a little and explain what we're talking about.

———

A conventional dynamical system has an explicit, pre-stated phase space. That is, there exists a simple, precise description of everything that the system can possibly do, and in some sense this description is known in advance. In addition, there is a fixed rule, or rules, that takes the current state of the system and transforms it into the next state. For example, if we are trying to understand the solar system, from a classical point of view, then the phase space comprises all possible positions and velocities for the planets, moons, and other bodies, and the rules are a combination of Newton's law of gravity and Newton's laws of motion.

Such a system is deterministic: in principle, the future is entirely determined by the present. The reasoning is straightforward. Start with the present state and work out what it will be one time-step into the future by applying the rules. But we can now consider that state as the new 'present' state, and apply the rule again to find out what the system will be doing two time-steps into the future. Repeat again, and we know what will happen after three time-steps. Repeat a billion times, and the future is determined for the next billion time-steps.

This mathematical phenomenon led the eighteenth-century mathematician Pierre Simon de Laplace to a vivid image of a 'vast intellect' that could predict the entire future of every particle in the universe, once it was furnished with an exact description of all those particles at *one* instant. Laplace was aware that performing such a computation was far too difficult to be practical, and he was also aware of the difficulty, indeed the impossibility, of observing the state of every particle at the same moment. Despite these problems, his image helped to create an optimistic attitude about the predictability of the universe. Or, more accurately, of small enough bits of it. And for several centuries, science made huge inroads into making such predictions feasible. Today, we can predict the motion of the solar system billions of years in advance, and we can even predict the weather (fairly accurately) *three whole days* in advance, which is amazing. Seriously. Weather is a lot less predictable than the solar system.

Laplace's hypothetical intellect was lampooned in Douglas Adams's *The Hitchhiker's Guide to the Galaxy* as Deep Thought, the supercomputer which took five million years to calculate the answer to the great question of life, the universe, and everything. The answer it got was 42. 'Deep Thought' is not so far away from 'Vast Intellect', although the name originates in the pornographic movie *Deep Throat*, whose title was the cover-name of a clandestine source in the Watergate scandal in which the presidency of Richard Nixon self-destructed (how soon people forget ...).

One reason why Adams was able to poke fun at Laplace's dream is that about forty years ago we learned that predicting the future of the universe, or even a small part of it, requires more than just a vast intellect. It requires absolutely exact initial data, correct to infinitely many decimal places. No error, however minuscule, can be tolerated. *None*. No marks for trying. Thanks to the phenomenon known as 'chaos', even the smallest error in determining the initial state of the universe can blow up exponentially fast, so that the predicted future quickly becomes wildly inaccurate. In practice, though, measuring anything to an accuracy of more than one part in a trillion, 12 decimal digits, is beyond the abilities of today's science. So, for instance, although we can indeed predict the motion of the solar system billions of years in advance, we can't predict it *correctly*. In fact, we have very little idea where Pluto will be, a hundred million years from now.

Ten million, on the other hand, is a cinch.

Chaos is just one of the practical reasons why it's generally impossible to predict the future (and get it right). Here we'll examine a rather different one: complexity. Chaos afflicts the prediction method, but complexity afflicts the rules. Chaos occurs because it is impossible to say in practice what the state of the system is, exactly. In a complex system, it may be impossible to say what the range of possible states of the system is, even approximately. Chaos throws a spanner in the works of the scientific prediction machine, but complexity turns that machine into a small cube of crumpled scrap metal.

We've already discussed the limitations of the Laplacian world-picture in the context of Kauffman's theory of autonomous agents expanding into the space of the adjacent possible. Now we'll take a closer look at how such expansions occur. We'll see that the Laplacian picture still has a role to play, but a less ambitious one.

A complex system consists of a number (usually large) of entities or agents, which interact with each other according to specific rules. This description makes it sound as though a complex system is just a dynamical system whose phase space has a huge number of dimensions, one or more per entity. This is correct, but the word 'just' is misleadingly dismissive. Dynamical systems with big phase spaces can do remarkable things, far more remarkable than what the solar system can do.

The new ingredient in complex systems is that the rules are 'local', stated on the level of the entities. In contrast, the interesting features of the system itself are global, stated on the level of the entire system. Even if we know the local rules for entities, it may not be possible – either in practice, or in principle – to deduce the dynamical rules of the system as a whole. The problem here is that the calculations involved may be intractable, either in the weak sense that they would take far too long to do, or in the strong sense that you can't actually do them at all.

Suppose, for example, that you wanted to use the laws of quantum mechanics to predict the behaviour of a cat. If you take the problem seriously, the way to do this is to write down the 'quantum wave-function' of every single subatomic particle in the cat. Having done this, you apply a mathematical rule known as Schrödinger's equation, which physicists tell us will predict the future state of the cat.*

However, no sensible physicist would attempt any such thing, because the wavefunction is far too complicated. The number of sub-atomic particles in a cat is enormous; even if you could measure their states precisely – which of course you can't do anyway – the universe does not contain a sheet of paper big enough to list all the numbers. So the calculation can't even get started, because in practical terms the present state of the cat is indescribable in the language of quantum wavefunctions. As for plugging the wavefunction into Schrödinger's equation, well, forget it.

Agreed, this is not a sensible way to model the behaviour of a cat. But it does make it clear that the usual physicists' rhetoric about quantum mechanics being 'fundamental' is at best true in a philosophical sense. It's not fundamental to our *understanding* of the cat, although it might be fundamental to the cat.

Despite these difficulties, cats generally manage to behave like cats, and in particular they discover their own futures by living them. Down on the philosophical level, again, this may be because the universe is a lot better at solving Schrödinger's equation than we are, and because it doesn't need a description of the quantum wavefunction of the cat: it's already got the cat, which *is* its own quantum wavefunction from

---

* Schrödinger pointed out that quantum mechanics often gives silly answers like 'the cat is half alive and half dead'. His intention was to dramatise the gap between a quantum-level description of reality and the world we actually live in, but most physicists missed the point and derived complicated explanations of why cats really are like that. And why the universe needs conscious observers to ensure that it con-tinues to exist. Only recently did they twig what Schrödinger was on about, and come up with the concept of 'decoherence', which shows that superpositions of quantum states rapidly change into single states unless they are protected from inter-action with the surrounding environment. And the universe doesn't need *us* to make it hold together, sorry. See *The Science of Discworld*, with a cameo appearance of Nanny Ogg's cat Greebo.

this point of view.

Let's accept that, even though it's rather likely that the universe does-n't propagate a cat into its future by applying anything that corresponds to Schrödinger's equation. The equation is a human model, not the reality. But even if Schrödinger's equation is what the universe 'really' does – and more so if it's not – there's no way that we limited humans can follow the 'calculation' step by step. There are too many steps. What interests us about cats occurs on the system level: things like purring, catching mice, drinking milk, getting stuck in the catflap. Schrödinger's equation doesn't help us understand those phenomena.

When the logical chain that leads from an entity-level description of a complex system to system-level behaviour is far too complicated for any human being to follow it, that behaviour is said to be an *emergent property* of the complex system, or just to be 'emergent'. A cat drinking milk is an emergent property of Schrödinger's equation applied to the subatomic particles that make up the cat. And the milk, and the saucer ... and the kitchen floor, and ...

One way to predict the future is to cheat. This method has many advantages. It works. You can test it, so that makes it *scientific*. Lots of people will believe the evidence of their own eyes, unaware that eyes tell lies and you'll never catch a competent charlatan in the act of cheating.

The wizards got Shakespeare right, aside – at a late stage – from the minor matter of sex. When it comes to a baby's sex, the Grand Master of Foretelling the Future was 'Prince Monolulu'. He was a West African who wore very impressive tribal gear and haunted (in a very material sense) the markets in the East End of London in the 1950s. Prince Monolulu would accost pregnant women with the cry 'I will tell you the sex of your baby, money back guarantee!' Many ladies fell for this ploy, and paid a shilling, then about a fiftieth of one week's wages.

Level One of the trick is that random guesses would guarantee the Prince 50 per cent of the money, but he was much more cunning than that. He improved the scheme to Level Two by writing the prophecy on a note, putting it into an envelope, and getting the sucker to sign across the seal. When it turned out that the anticipated John was really

Joan, or Joan was John, the few who bothered to return to reclaim their money found that, on opening their envelope, it contained a correct prediction. They didn't get their money back, because Prince Monolulu insisted that what was in the envelope was what he had originally told them; the sucker must have remembered it wrong. In reality, the envelope always contained the opposite prediction to the verbal one.

History is a complex system; its entities are people, its rules of interaction are the complicated ways in which human beings behave towards each other. We don't know enough sociology to write down effective rules at this entity level. But even if we did, the system-level phenomena, and the system-level rules that govern them, would almost certainly be emergent properties. So the rule that propagates the state of the entire system one step into the future is not something we can write down. It is an emergent dynamic.

When the system-level dynamic is emergent, then even the system itself does not 'know' where it is going. The only way to find out is to let the system run and see what happens. You have to allow the system to make up its own future as it goes along. In principle only one future is possible, but there is no short cut that lets you predict what will happen before the system itself gets there and we all find out. This behaviour is typical of complex systems with emergent dynamics. In particular, it is typical of human history and of biological evolution. And cats.

Biologists learned long ago not to trust evolutionary explanations in which the evolving organisms 'knew' what they were trying to achieve. Explanations like 'the elephant evolved a long trunk in order to suck up water without bending down'. The objectionable item here is not the reason why the elephant's trunk is long (though, of course, that can be debated): it is the phrase 'in order to'. This endows elephants with evolutionary prescience, and suggests (wrongly) that they can somehow choose the direction in which they evolve. All this is obvious nonsense, so it's not sensible to have a theory that attributes purpose to elephant evolution.

Unfortunately, a dynamic looks remarkably like purpose. If elephant evolution follows a dynamic, then it looks as if the end result is pre-determined, in which case the system 'knows' in advance what it ought to be doing. The individual elephants need not be conscious of their objective, but the *system* in some sense has to be. That would be a good argument against a dynamic description if the evolutionary dynamic for elephants was something we could prescribe ahead of time. However, if that dynamic is emergent, then the system itself, along with the elephants, can find out where it's headed only by going there and discovering where it gets to.

The same goes for history. Being able to put a name to a historical period only after it's happened looks remarkably like what you'd observe if there is a historical dynamic, but it is emergent.

This far into the discussion, it may seem that an emergent dynamic is no better than no dynamic at all. Our task now is to convince you that this is not so. The reason is that although an emergent dynamic cannot be deduced, in complete logical detail, from entity-level rules, *it is still a dynamic*. It has its own patterns and regularities, and it may be possible to work with those directly.

Exactly this is going on when a historian says something like 'Croesus the Unprepared was a rich but weak king who never main-tained a sufficiently large army. It was therefore inevitable that his kingdom would be overrun by the neighbouring Pictogoths, and his treasury would be plundered'. This kind of story proposes a system-level rule, a historical pattern, which can sometimes be compelling. We can question how scientific such stories are, because it is always easy to be wise after the event. But in this case the story generalises: rich weak kings are asking to be invaded by mean, poor barbarians. And that's a prediction, wisdom before the event, and as such it is sci-entifically testable.*

The stories that evolutionary biologists tell are of the same kind, and they become science when they stop being Just-So Stories, justifications after the event, and become general principles that make predictions. These predictions are of a limited kind: 'in *these* circumstances expect

---

* Discworld runs this far more sensibly. Heroes will have adventures.

this behaviour'. They are not predictions of the type 'On Tuesday at 7.43pm the first elephant trunk will evolve'. But this is what 'prediction' means in science: saying ahead of time that under certain conditions, certain things will happen. You don't have to predict the timing of the experiment.

An evolutionary example of this kind of pattern can be found in the co-evolution of 'creodonts', big cats like sabretooth tigers, and their 'titanothere' prey – large-hoofed mammals, often with huge horns. When it comes to improving performance for the big cats, the line of least resistance is to develop bigger teeth. Faced with that, the best response for the prey is to develop thicker skins and bigger horns. An evolutionary arms race now becomes pretty much unavoidable: the cats get bigger and bigger teeth, and the prey respond with thicker and thicker skins ... to which the cats' only response is even bigger teeth ... and so it goes. An evolutionary arms race sets in, with both species trapped in a single strategy. The end result is that the cats' teeth get so enormous that the poor animals can hardly move their heads, while the titanotheres' skins, and multiple horns on nose and brow, and associated musculature, get so heavy that they find trouble dragging themselves across the plains. Both species promptly die out.

This creodont-titanothere arms race has happened at least five times in evolutionary history, taking about five million years to run its course on each occasion. It is a striking example of an emergent pattern, and the fact that it plays out in exactly the same way over and over again confirms that there really is an underlying dynamic. In all likelihood it would be happening again, now, except for the arrival of humans, who have clobbered both the big cats and their slow prey.

Notice that we've been calling these system-level patterns 'stories', and so they are. They have a narrative, a consistent internal logic; they have a beginning and an end. They are stories because they cannot be 'reduced' to an entity-level description; that would be more like an interminable soap opera. 'Well, this electron bumped into that electron and the two of them got together and emitted a photon ...' repeated, with slight variations, a truly inconceivable number of times.

One of the central questions about emergent dynamics is: what would happen if we ran the system again, in slightly different circumstances? Would the same patterns emerge, or would we see something completely different? If European history in the early twentieth century was rerun, but without Adolf Hitler, would World War II have happened anyway, by a different route? Or would it all have been sweetness and light? Historically, this is a crucial question. There is no doubting that Hitler was instrumental in starting World War II; the deeper question here is whether he was a product of the politics of the time, and in his absence someone else would have done much the same, or whether it was Hitler who moulded history and created a war when otherwise nothing would have happened.

At risk of being controversial, we are inclined to the view that World War II was a pretty much inevitable consequence of the political situation in the 1930s, with Germany saddled with huge reparations for World War I, the trains not running on time … and Hitler was merely the medium through which the national will to war was expressed. But it's not the answer that concerns us here: it is the nature of the question. It is a 'what if' question, and it is about historical phase space. It does not ask what happened; it asks what might have happened instead.

This point is well understood on Discworld. In *Lords and Ladies* we find the following passage:

> There are indeed such things as parallel universes, although parallel is hardly the right word – universes swoop and spiral around one another like some mad weaving machine or a squadron of Yossarians* with middle-ear trouble.
>
> And they branch. But, and this is important, not all the time. The universe doesn't much care if you tread on a butterfly. There are plenty more butterflies. Gods might note the fall of a sparrow but they don't make any effort to catch them.
>
> Shoot the dictator and prevent the war? But the dictator is merely the tip of the whole festering boil of social pus from which dictators

---

* Recall that Yossarian is a pilot in Joseph Heller's *Catch-22*.

emerge; shoot one and there'll be another one along in a minute. Shoot him too? Why not shoot everyone and invade Poland? In fifty years', thirty years', ten years' time the world will be very nearly back on its old course. History always has a great weight of inertia.

Almost always ...

At circle time, when the walls between this and that are thinner, when there are all sorts of strange leakages ... Ah, then choices are made, then the universe can be sent careening down a different leg of the well-known Trousers of Time.

This kind of question can be asked of any dynamical system, emergent or not; but it takes on a special aspect when the dynamic 'makes itself up as it goes along'. In a rerun, would it make up the same thing? Would it tell the same story? If so, that story is robust: it has a degree of inevitability, not just in some particular run of history, but in all of them.

Science fiction writers explore historical phase space in 'alternate* universe' stories, where one historical event is changed and the author develops possible consequences. Philip K. Dick's *The Man in the High Castle* explores a history in which Germany won World War II. Harry Harrison's *West of Eden* trilogy explores a world in which the K/T meteorite missed and the dinosaurs survived. Science writers also ask about historical phase space, especially in the context of evolution. The most celebrated example is Stephen Jay Gould's *Wonderful Life*, which asks whether humans would arise again on Earth if evolution were to be run again. His answer, 'no', rests on a very literal interpretation of 'human'. Harrison's answer in *West of Eden* is that intelligent mosasaurs – contemporaries of the dinosaurs that had returned to the sea – would evolve, and play the same role on the evolutionary stage that humans have played in this world. (For plot reasons he also has genuine humans in his alternate universe, but the Yilané, the smart mosasaur descendants, were there first.)

Where Gould sees divergence and massive changes brought about

---

* We use this word because it's standard in science fiction, but UK English would require 'alternative'.

by chance events, Harrison sees convergence: same play, different actors. To Gould, a change of actor is significant; to Harrison, what matters is the play. Both have good arguments to present, but the main point is that they are tackling different questions.

A second way in which science fiction writers explore alternative historical tracks is through the time travel story, and this brings us back to the wizards of Unseen University and their battle against the elves. There are two kinds of time travel story. In the first kind, the protagonists mainly use their ability to travel in time as a way of observing the past or future; a good example is the first significant time travel novel, H.G. Wells's *The Time Machine* of 1895. The time machine is a vehicle for Wells to discuss the future of humanity, but his Time Traveller makes no real effort to *change* history. In contrast, the narrative theme of Robert Silverberg's 1969 novel *Up the Line* is the paradoxes that arise if it is possible to travel into the past and change it. In this story, the Time Service does not set out to change the past; on the contrary, its prime objective is to preserve the past and avoid paradoxes, despite the activities of observers from the future, who are cataloguing the past by visiting it and seeing what actually happened.

The classic time travel paradox is 'what if I went back and killed my grandfather?' The logic of the situation, of course, is that with granddad dead, you wouldn't have been born, so you wouldn't be able to go back and kill him, so he'd have lived, so you would have been born ... All attempts to resolve this self-contradictory causal loop are cheats: perhaps granddad dies, but you get born anyway with different grandparents, but then it wasn't really granddad that you killed. In the 'many worlds' interpretation of quantum mechanics, the causal logic of the universe holds together provided the grandfather that gets killed was in a different parallel universe from that of the killer. But then he wasn't your real granddad, either, just a parallel version in some other universe.

A slightly more subtle time paradox is the Cumulative Audience Paradox. If people in the future have access to time machines, then they are bound to want to go back and witness all of the great historical events, like the crucifixion. But we know, from existing descriptions of these events, that they did not happen in front of crowds of thousands of visitors from the future. So where were they? This is a temporal analogue of the Fermi Paradox* about intelligent aliens: if they're all over the galaxy, then why aren't they here? Why haven't they visited us? Other time paradoxes are used as essential plot elements in Robert A. Heinlein's short stories 'By his bootstraps' and 'All you zombies'. In the latter, a time-traveller manages to be his own father, son, and – via a sex change – mother. When asked where he comes from, he replies that he knows exactly where he comes from. The big puzzle is: where does everybody *else* come from? This idea is taken to serious extremes by David Gerrold in *The Man Who Folded Himself*.

Over the last few decades, serious physicists have started thinking about the possibility of time travel and the resolution of any associated paradoxes. Their work is a tribute to narrative imperative on Roundworld. The reason they are asking such questions is no doubt that as children they read stories like those of Wells, Silverberg, Heinlein and Gerrold. When they became professional physicists, the stories bubbled up from their subconscious, and they began to take the idea seriously – not as a practical engineering issue, but as a theoretical challenge.

Do the laws of physics permit time travel, or not? You'd expect the answer to be 'no', but the remarkable consequence of the theorists' research is that it is 'yes'. A working time machine is still a long way off, and it may be that we're missing some basic physical principle that would change the answer to 'no', but the fact is that today's accepted frontier physics does not forbid time travel. It even offers a few scenarios in which it could occur.

The context for such research is general relativity, in which the continuum of space and time can be distorted by gravity. Or, more accurately, in which gravity is caused by such distortions, 'curved spacetime'. In place of a time machine, the physicists look for a 'closed timelike curve'. Such a curve corresponds to an object that travels into the future and ends up in its own past, and so becomes trapped in a

---

* Named after the physicist Enrico Fermi. See *Evolving the Alien* by Jack Cohen and Ian Stewart.

closed 'time loop'.

The best known way to generate a closed timelike curve is to use a wormhole. A wormhole is a short-cut through space, obtained by fusing a Black Hole to its time-reversal, a White Hole. Just as Black Holes suck in anything that comes near them, White Holes spit things out. A wormhole sucks things in at its black end and spits them out at its white end. Of itself, a wormhole is more a matter-transmitter than a time machine, but it becomes a time machine when allied to the famous Twin Paradox. In relativity, time slows down for objects moving at very high speeds. So if one member of a pair of twins heads out to a distant star at very high speed, and then returns, she will have aged less than the other twin who stayed at home. Suppose that the travelling twin takes with her the white end of a wormhole, while her sister keeps the black end. Then when the travelling twin returns, the white end is younger than the black end: the exit from the wormhole lies in the past of the entrance. So anything that is sucked into the black end is spat out in its own past. Because the white end is now right next to the black one – the twin has come back home – the object can hop across to the Black Hole and go round and round this closed loop in spacetime, tracing a closed timelike curve.

There are practical problems in making such a gadget, the main one(!) being that the wormhole will collapse too quickly for an object to pass through it, unless it is held open by threading 'exotic matter' with negative energy through it. Nonetheless, none of this is forbidden by the current laws of physics. So what of the paradoxes? It turns out that the laws of physics forbid genuine paradoxes, although they permit many apparent paradoxes. A useful technique for understanding the difference is known as a Feynman diagram, which is a picture of the motion of an object (usually a particle) in space and time.

For example, here is an apparent time travel paradox. A man is imprisoned in a concrete cell, locked from the outside, with no food, no water and no possibility of escape. As he sits in a corner in despair, waiting for death, the door opens. The person who has opened it is … himself. He has returned in a time machine from the future. But how (the paradox) did he get to the future in the first place? Well, a kind person opened the door and set him free …

There seems to be something very odd about the causality in this story, but the corresponding Feynman diagram shows that it violates none of the laws of physics. First, the man follows a space-time path that puts him inside the cell and then removes him from it through an opened door. This time-line continues into his future until he encounters a time machine. Then the time-line reverses direction, heading into the past, until he encounters a locked cell. He opens it, and his time-line reverses again, propelling him into his own future. So the man follows a single zig-zag path through time, and at every step the laws of physics hold good. Provided his time machine violates no physical law, of course.

If you try to 'explain' the grandfather paradox by this method, it doesn't work. The time-line leading from grandfather to killer is severed when the killer returns; there is no consistent scenario, even in a Feynman diagram. So some stories of time travel are consistent with the laws of physics, and have their own kind of causal logic, albeit twisted; but other equally plausible stories are inconsistent with the laws of physics. You can rescue the Grandfather Paradox by assuming that changing the past in a logically inconsistent way switches you into a different alternate universe – say a quantum-mechanical parallel world. But then it wasn't your grandfather that you killed, but the grandfather of an alternate you. So this 'resolution' of the Grandfather Paradox is a cheat.

Faced with all this, the way that the wizards handle the complications of time travel seems quite reasonable!

TWENTY-NINE
ALL THE GLOBE'S A THEATRE

 THE ELVES DID NOT SPEND A LOT OF TIME in serious thought. They could control people who could do the thinking for them. They didn't play music, they did not paint, they never carved stone or wood. Control was the talent, and it was the only one they had ever needed.

Nevertheless, there were ones who had survived for many thousands of years, and while they had no great intelligence they had accumulated that mass of observations, experience, cynicism and memory that can pass for wisdom among people who don't know any better. One of the wisest things they did was not read.

They had found some clerks to read the play.

They listened.

Then, when it was over, the Queen said: 'And the wizards have been showing great interest in this man?'

'Yes, your majesty,' said one of the old ones.

The Queen frowned. 'This ... *play* is ... good. It treats us ... kindly. We are firm but fair with mortals. We offer rewards to those who deal well with us. Our beauty is satisfactorily referred to. Our ... issues with our husband are treated more romantically than I would like, but, nevertheless ... it is positive, it enhances us, it places us yet more firmly in the human world. One of the wizards was actually carrying this.'

One of the senior elves cleared its throat. 'Our grip is loosening, your majesty. Humanity is becoming more, shall we say, questioning?'

The Queen shot it a glance. But it was older than many Queens, and did not step back.

'You think it will do us harm? Is it a plot against us?'

The senior elves looked at one another. The main reason that they thought it was a plot was that they were predisposed to see plots. In the court of Faerie, an inability to see it coming meant that it took you by the throat.

'We think it may be,' one said at last.

'How? In what way?'

'We know the wizards have been seen in the company of the author,' said the elf.

'Then perhaps they are endeavouring to *stop* him writing the play, have you thought of that?' snapped the Queen. 'Can you see *any* way in which those words harm us?'

'We are agreed that we cannot … nevertheless, we have a sense that in some way—'

'It is so simple! At last we are done some real honour and the wizards will try to stop it! Are you so stupid that you cannot see that?'

Her long dress swirled as she turned on her heel. 'It *will* happen,' she said. 'I will see to it!'

The senior elves filed out, not looking at her face. They knew those moods.

On the stairs one said to the others: 'Purely out of interest … can any of *us* put a girdle around the Earth in three minutes?'

'That would be a very big girdle,' said an elf.

'And would *you* wish to be called Peaseblossom?'

The eyes of the old elf were grey, flecked with silver. They had seen horrible things under many suns, and in most cases had enjoyed them. Humans were a valuable crop, the elf conceded. There had never been a species like it for depth of awe, terror and superstition. No other species could create such monsters in its heads. But sometimes, it considered, they were not worth the effort.

'I think not,' it said.

'Well, now, Will – do you mind if I call you Will? Oh, Dean, fetch Will another pint of this really unpleasant ale, will you? Now … where was I … oh yes, I really enjoyed that play of yours. Magnificent, I thought!' Ridcully beamed. Around him, the inn hummed with life.

Will tried to focus. 'Which one wa*f* that, good *f*ir?' he said.

Ridcully's smile remained fixed, but began to unravel around the edges. He was never one for unnecessary reading.

'The one with the king in it,' he said, aiming for safety.

On the other side of the table Rincewind did some desperate pantomime.

'The rabbit,' said Ridcully. 'The rat. The ferret. Sounds like … hat. Rat. Rodent. Thing with teeth.'

Rincewind gave up, leaned across and whispered.

'Something about the shrew,' said Ridcully. Rincewind whispered a little harder.

'The one about the tame shrew. The man married a shrew. A shrewish woman. Not a real shrew, obviously, haha. No one would marry a real shrew. It would be a completely foolish idea.'

Will blinked. He was not, as an actor and a writer, averse to alcohol bought by other people, and these people were being very good hosts. It was just that they seemed to be completely deranged.

'Er … I thank you,' he said. He was aware of being stared at, and also of a strange but not unpleasant animal smell. He turned on the bench and was rewarded with a grin. It occupied all the space between a deep hood and a jerkin. There were a couple of brown eyes, too, but it was the grin his gaze kept coming back to.

The Librarian raised his tankard and gave Will a friendly nod. This caused the grin to get bigger.

'Now I'm sure you hear this all the time,' said Ridcully, slapping Will so hard on the back that his drink slopped, 'but we've got an idea for you. Dean, more ale all round, eh? It really is very weak stuff. Yes, an idea.' He poked Will in the chest. 'Too many kings, that's the trouble. What the public wants now, what puts bums on seats—'

'Feet,' said Rincewind.

'What?'

'Bums on feet, Archchancellor. It's mostly standing room in the theatre.'

'Feet, then. Bums, anyway. Thank you, Dean. Cheers.' Ridcully wiped his mouth delicately and turned his attention again to Will, who tried to avoid the prodding finger.

'Bums on, haha, feets,' he said, and blinked. 'Funny thing, funny thing, something similar happened to us, 'smatterofact, few years ago, Midsummer's Eve, these chaps were going to put on a play thingy for the king, next thing, elves all over the place, haha. Why, yes, Runes, I'll have another if you're paying, it's far too sweet to be a serious drink. Where was I? Ah. Elves. What you've got to do, what you've gotta do ... is ... why aren't you writing this down?'

In the morning Rincewind opened his eyes at the fourth attempt and with the assistance of both hands. There was a moment of brain lag, where the little wheels spun happily with no work to do, and then big horrible machinery cut in.

'Whg d'hl der ...' he said, and then got control of his mouth as well.

Bits of last night crept out of hiding to do their treacherous dance before his eyes. He groaned.

'We couldn't have done that, could we?' he muttered.

And memory said: that was only the start ...

Rincewind sat up and waited until the world stopped moving.

He'd been on the floor in the library. The other wizards lay scattered around the room or sprawled across piles of books. The air smelled of beer.

A veil will be drawn over the following half an hour, and lifted to find the wizards sitting around the table.

'It must've been the pork scratchings,' said the Dean.

'I don't remember any pork scratchings,' muttered Ponder.

'Something crunchy, anyway. They may have been moving about.'

'There's no doubt in my mind that it was caused by all this travelling we've been doing,' said Ridcully. 'That sort of thing must take a terrible toll on the system. We've been concentrating so hard, d'yer see, that the moment we relaxed the strain we just unwound, like a big spring.'

The wizards brightened up. Rascally drunkenness was too much of an embarrassment to men who could sit through an entire meal at the UU high table, but time sickness ... yes, that had a certain cachet. They could live with time sickness although, at the moment, they were wishing they didn't have to.

'That's right!' said the Lecturer in Recent Runes. 'It wasn't the fight!'

'And it couldn't possibly have been the carousing, which was really quite moderate by our standards,' said the Dean.

'In fact we didn't get drunk at all!' said the Chair of Indefinite Studies, brightly.

Unfortunately, Rincewind's memory was literally treacherous. It worked perfectly.

'So, then,' he said, wishing that he didn't have to, 'we didn't tell Will all that stuff?'

'What stuff?' said Ridcully.

'All about our magical library, for one thing. And you kept saying "Here's a good one, I bet you can use this" and you told him about those witches up in Lancre and how they got the new king on the throne, and that time the elves broke through, and how the Selachii and the Venturi families are always fighting—'

'We did?' said Ridcully.

'Yes. And about the countries we've visited. Lots of things.'

'Why didn't someone stop me?'

'The Dean did try. That's when you hit him with the Chair of Indefinite Studies, I think.'

The wizards sat in ale-smelling gloom.

'Should we have another try?' said the Lecturer in Recent Runes.

'What, and tell him to forget it all?' said Ridcully. 'Talk sense, man.'

'Perhaps we could go back in time and stop ourselves telling—'

'Don't say that! No more of that!' snapped the Archchancellor.

Rincewind pulled a copy of the play towards him. The wizards froze.

'Go on,' said Ridcully. 'Tell us the worst. What did he write?'

Rincewind opened the book and read a couple of lines at random:

> *You spotted snakes, with double tongue;*
> *Thorny hedgehogs, be not—*

'No, no, no,' muttered the Dean, his head in his hands. 'Please tell me no one sang him the Hedgehog Song …'

Rincewind's lips moved as he read on. He turned over a few pages.

He flicked back to the beginning.

'It's all here,' he said. 'Same rather bad jokes, same unbelievable confusions, everything! Just as it was before! But it's going to happen here!'

The wizards looked at one another and dared to share a smug expression.

'Ah well, there we are then,' said Ridcully, sitting back. 'Job done.'

Rincewind turned some more pages. His recollections of the night were not coherent, but even a genius couldn't have made sense out of a bunch of drunken wizards all talking at once.

'Hex?' he said.

The crystal ball said: 'Yes?'

'Will this play be performed in this world?'

'That is the intention,' said the voice of Hex.

'And then what will happen?'

Hex told them, and added: 'That is one outcome.'

'Just a moment,' said Ponder Stibbons. 'There's more than one outcome?'

'Certainly. The play may not take place. Phase space contains a broadsheet account of a disruption of the first performance, followed by a fire in which a number of people died. Subsequently the theatres were closed and the playwright died during a riot. He was struck by a pike.'

'You mean a halberd, of course,' said Ridcully.

'A pike,' Hex repeated. 'A fishmonger was involved.'

'What happened to civilisation?'

Hex was silent for a moment, and then said: 'Humanity failed by three years to leave the planet.'

THIRTY
LIES TO HUMANS

 PLEASE TELL ME NO ONE SANG him the Hedgehog Song …

The Hedgehog Song, a Discworld ditty in the general tradition of Eskimo Nell, first made its appearance in *Wyrd Sisters* with its haunting refrain 'The hedgehog can never be buggered at all'. The wizards have wielded the power of story with a vengeance. They have used it to prime their secret weapon, Shakespeare, and are convinced that he will prove more effective than a MIRVed ICBM. But before he's launched, they've very properly started to worry about collateral damage: possible cultural contamination by the Hedgehog Song.

It is a consequence only marginally less dire than eternal elf-infestation, but on the whole, preferable.

In the real Roundworld, the power of story is just as great as it is in the fictional counterpart. Stories have power because we have minds, and we have minds because stories have power. It's a complicity, and all that remains is to unwrap it.

As we do so, bear in mind that Discworld and Roundworld are not so much different as complementary. Each, in its own estimation at least, gave birth to the other. On Roundworld, the Disc is seen as fantasy, the invention of an agile mind; Discworld is a series of stories (amazingly successful) along with ceramic models, computer games and cassette tapes. Discworld runs on magic, and on narrative imperative. Things happen on Discworld because people assume they will, and because some things have to happen to complete the story. From the standpoint of Roundworld, Discworld is a Roundworld invention.

The Discworld view is similar, but inverted. The wizards of Unseen University *know* that Roundworld is merely a Discworld creation, an unanticipated spin-off from an all-too-successful attempt to split the thaum and create the first self-sustaining magical chain reaction. They know this because they were there when it happened. Roundworld was deliberately created to keep magic out. Surprisingly, the magic-free vacuum acquired its own regulatory principle. Rules. Things happen on Roundworld because they are consequences of the rules. However, it is astonishingly difficult to look at the rules and understand what their consequences will be. Those consequences are emergent. The wizards discovered this to their cost, as every attempt to do something straightforward in Roundworld – like creating life or jump-starting extelligence – went seriously awry.

These two worldviews are not mutually contradictory, for they are worldviews of two different worlds. Yet, thanks to the interconnectedness of L-space, each world illuminates the other.

The strange duality between Roundworld and Discworld parallels another: the duality between Mind and Matter. When Mind came to Roundworld, a very remarkable change occurred. Narrative imperative appeared in Roundworld. Magic came into existence. And elves, and vampires, and myth, and gods. Characteristically, all of these things came into being in an indirect and offbeat way, like the relationship between rules and consequences. Things didn't exactly *happen* because of the power of story. Instead, the power of story made minds *try* to make the things in the story happen. The attempts were not always successful, but even when they failed, Roundworld was usually changed.

Narrative imperative arrived on Roundworld like a small god, and grew in stature according to human belief. When a million human beings all believe the same story, and all try to make it come true, their combined weight can compensate for their individual ineffectiveness.

There is no science in Discworld, only magic and narrativium. So the wizards put science into Discworld in the form of the Roundworld Project, as detailed in *The Science of Discworld*. With elegant symmetry, there was no magic or narrativium in Roundworld, so humans put them there, in the form of story.

Before narrative imperative can exist, there has to be narrative, and that's where Mind proved decisive. The imperative followed hard on the heels of the narrative, and the two complicitly co-evolved, for as soon as there was a story, there was someone who wanted to make it come true. Nonetheless, the story beat the compulsion by a nose.

What makes humans different from all other creatures on the planet is not language, or mathematics, or science. It is not religion, or art, or politics, either. All of those things are mere side effects of the invention of story. Now it might seem that without language there can be no stories, but that is an illusion, brought about by our current obsession with recording stories as words on paper. Before there was a word for 'elephant' it was possible to point at an elephant and make evocative gestures, to draw an elephant on the cave wall and add spears flying towards it, or to mould a model of an elephant from clay and act out a hunting scene. The story was as clear as day, and an elephant-hunt would follow hard on its heels.

We are not *Homo sapiens*, Wise Man. We are the third chimpanzee. What distinguishes us from the ordinary chimpanzee *Pan troglodytes*, and the bonobo chimpanzee *Pan paniscus*, is something far more subtle than our enormous brain, three times as large as theirs in proportion to body weight. It is what that brain makes possible. And the most significant contribution that our large brain made to our approach to the universe was to endow us with the power of story. We are *Pan narrans*, the storytelling ape.

Even today, five million years since we and the other two species of chimpanzee went our separate evolutionary ways, we still use stories to run our lives. Every morning we buy a newspaper to find out, so we tell ourselves, what is happening in the world. But most things that are happening in the world, even rather important ones, never make it into the papers. Why not? Because newspapers are written by journalists, and every journalist learned at their mother's knee that what grabs newspaper readers is a *story*. Events with zero significance for the planet, such as a movie star's broken marriage, are stories. Events that matter a great deal, such as the use of chlorofluorocarbons (CFCs) as propellants in aerosol cans of shaving-cream, are not stories. Yes, they can become stories, and in this case do when we discover that those selfsame CFCs are destroying the ozone layer; we even have a title for the story, The Ozone Hole. But nobody knew or recognised there was a story when shops first started selling aerosol cans, even though that was the decisive event.

Religions have always recognised the power of a good story. Miracles run better at the box-office than mundane good actions. Helping an old lady across the road isn't much of a story, but raising the dead most certainly is. Science is riddled with stories. In fact, if you can't tell a convincing story about your research, nobody will let you publish it. And even if they did, nobody else would understand it. Newton's laws of motion are simple little stories about what happens to lumps of matter when they are given a push – stories only a little more precise than 'if you keep pushing, it will go faster and faster'. And 'Everything moves in circles', as Ponder would insist.

Why are we so wedded to stories? Our minds are too limited to grasp the universe for what it is. We're very small creatures in a very big world, and there is no way that we could possibly represent that world in full, intricate detail inside our own heads. Instead, we operate with simplified representations of limited parts of the universe. We find simple models that correspond closely to reality extremely attractive. Their simplicity makes them easy to comprehend, but that's not much use unless they also *work*. When we reduce a complex universe to a simple principle, be it The Will of God or Schrödinger's Equation, we feel that we've really accomplished something. Our models are stories, and conversely, stories are models of a more complex reality. Our brains fill in the complexity automatically. The story says 'dog' and we immediately have a mental picture of the beast: a big, bumbling Labrador with a tail like a steam-hammer, tongue lolling, ears flopping.* Just as our visual system fills in the blind spot.

We learn to appreciate stories as children. The child's mind is quick and powerful, but uncontrolled and unsophisticated. Stories appeal to it, and adults rapidly discovered that a story can put an idea into a child's head like nothing else can. Stories are easy to remember, both

---

* It then comes as quite a jolt when we discover that the animal is a chihuahua.

for teller and listener. As that child grows to adulthood, the love of stories remains. An adult has to be able to tell stories to the next generation of children, or the culture does not propagate. And an adult needs to be able to tell stories to other adults, such as their boss or their mate, because stories have a clarity of structure that does not exist in the messiness of the real world. Stories always make sense: that's why Discworld is so much more convincing than Roundworld.

Our minds make stories, and stories make our minds. Each culture's Make-a-Human kit is built from stories, and maintained by stories. A story can be a rule for living according to one's culture, a useful survival trick, a clue to the grandeur of the universe, or a mental hypothesis about what might happen if we pursue a particular course. Stories map out the phase space of existence.

Some stories are just entertainment, but even those usually have a hidden message on a deeper, possibly more earthy, level – as with Rumpelstiltskin. Some stories are Worlds of If, a way for minds to try out hypothetical choices and imagine their consequences. Word-play in the Nest of the Mind. And some of those stories have such a compelling logic that narrative imperative takes over, and they transmute into plans. A plan is a story together with the intention of making it come true.

Inside Roundworld, as it sits in its glass globe within the confining walls of the library of Unseen University, our story is coming to its climax. Will Shakespeare has written a play (it is, of course, *A Midsummer Night's Dream*), a play that the elves believe will consolidate their power over human minds. The narrative of this play has collided with Rincewind's mental model of what he wants to do, and the flying sparks have ignited a plot. How will it all end? That is one of the compulsive aspects of a story. You'll just have to wait and see.

We have seen how history unfolds an emergent dynamic, so that even though everything is following rigid rules, even history itself has to wait and see how it all turned out. Yes, everything is following the rules, but there is no short cut that will take you to the destination before the rules themselves get there. History is not a story that exists in a book, the fatalistic 'it is written'. It is a story that makes itself up as it goes along, like a story that someone is reading and you are listening to. It is *being* written ...

Philosophically, there ought to be a big difference between a story that is already written, and one that is being created word by word as you read it. The one is a story whose every sentence is predetermined; not only can there be only one possible outcome, but the outcome is already 'known'. The other is a story whose next sentence does not yet exist, whose ending in unknown even to the storyteller. You are reading the first kind of story, but while we were writing it, it was the second kind of story. In fact, it started out as a totally different story, but we never wrote that one at all. The philosophers realised long ago that it is no easy matter to determine which kind of story fits our world. If we had the ability to run the world again, we might discover that it does different things on the second occasion, and if so, the history of the universe would be a story that unfolds as it goes, not one already committed to paper.

But this doesn't look like a feasible experiment.

Our fascination with stories lays us open to a variety of errors in our relationship with the outside world. The rapid spread of rumours, for instance, is a tribute to how our love of a juicy story overcomes our critical faculties. The mechanism is precisely the one that the scientific method tries very hard to protect us against: believing something because you want it to be true. Or, for some rumours, because you fear it could be true. A rumour is one example of a more general concept, introduced in 1976 by Dawkins in *The Selfish Gene*. He came up with this notion in order to be able to discuss an evolutionary system that was different from the Darwinian evolution of organisms. It is the *meme*. The associated subject of 'memetics' is science's attempt to comprehend the power of story.

The word 'meme' was coined by deliberate analogy with 'gene', and 'memetics' with 'genetics'. Genes are passed from one generation of organisms to the next; memes are passed from one human mind to another human mind. A meme is an idea that is so attractive to human minds that they want to pass it on to others. The song 'Happy Birthday to You' is a highly successful meme; so, for a long time, was Communism, though that was a complicated system of ideas, a *memeplex*. Ideas exist as some cryptic pattern of activity in brains, so brains, and their associated minds, provide an environment in which memes can exist and propagate. Indeed, replicate, for when you teach a child to sing 'Happy Birthday to You', you don't forget the song yourself. The Hedgehog Song is an equally successful Discworld meme.

As the home computer spread across the globe, and became inextricably wired into the Internet's extelligence, an environment was created that gave birth to an insidious silicon-based form of meme: the computer virus. All viruses so far seem to have been written deliberately by humans, although at least one turned out to be a far more successful replicator than its designer had intended, thanks to a programming error. 'Artificial life' simulations using evolving computer programs are often run inside a 'shell' that isolates them from the outside world, because of the unlikely but possible evolution of a really nasty computer virus. The world's computer network is certainly complex enough to evolve its own viruses, given enough time.

Memes are mind-viruses.

In *The Meme Machine*, Susan Blackmore says that 'Memes spread themselves around indiscriminately without regard to whether they are useful, neutral, or positively harmful to us.' The song 'Happy Birthday to You' is mostly harmless, although it is just about possible to see it as an insidious piece of propaganda for global commerce if you're that way inclined. Advertising is a conscious attempt to unleash memes; a successful advertising campaign starts to build its own momentum as it spreads by word of mouth as well as overt TV or newspaper ads. Some advertising is beneficial (Oxfam, say) and some is manifestly harmful (tobacco). In fact, many memes are harmful, but still propagate very effectively: among them are the chain-letter and its financial analogue, pyramid selling. Just as DNA propagates without having any conscious intentions of its own, so memes replicate without having conscious objectives. The people who set the memes loose may have had overt intentions, but the memes themselves don't. Those that perform well, leading human minds to pass them on in quantity, thrive; those that do not, die out, or at best live on as small, isolated pockets of infection. The spread of a meme is much like the spread of a disease. And just as you can protect yourself against some diseases, by taking the right precautions, you can also protect yourself against becoming infected with a meme. The ability to think critically, and to question statements that rest on authority instead of evidence, are quite effective defences.

This is our message to you. You need not be a victim of the power of story, like Vorbis the Quisitor, smitten by an earthbound tortoise, the Wrath of Om. You can be a Granny Weatherwax, sailing through story-space like a master navigator, attuned to every breath of narrative wind (and a lot of it is, mark you), tacking against the gale like a maverick, avoiding the Shoals of Dogma and the Scylla and Charybdis of Indecision …

Sorry, we got carried away. What we mean is: if you understand the power of story, and learn to detect abuses of it, you might actually deserve the appellation *Homo sapiens*.

Blackmore's book argues that many aspects of human nature are explained much better by memetics, the mechanisms whereby memes exist and propagate, than by any existing rival theory. In our terminology, memetics illuminates the complicity between intelligence and extelligence, between the individual mind and the culture of which it is but one tiny part. Some critics counter that the memeticists can't even say what the basic unit of a meme is. For example, are the first four notes of Beethoven's Fifth Symphony (dah-dah-da DUM) a meme, or is the meme really the whole symphony? Both replicate successfully: the second in the minds of music-lovers, the first in a weird variety of minds.

However, this kind of criticism never carries much weight when a new theory is being developed. Not that this stops the critics, of course. By the time a scientific theory can 'define' its concepts with complete precision, it's dead. Very few concepts can actually be defined completely: not even something like 'alive'. What, precisely, does 'tall' mean? 'Rich'? 'Wet'? 'Convincing'? Let alone 'slood'. If it comes to the crunch, the basic unit of *genetics* has not been defined in any convincing way, either. Is it a DNA base? A DNA sequence that codes for proteins, a 'gene' in the most limited sense? A DNA sequence with a known function – a 'gene' in its broadest sense? A chromosome? An entire genome? Does it have to exist inside an organism? Most DNA in the world contributes nothing genetic to the future: there's DNA in dead skin flakes, falling leaves, rotting logs …

Dawkins's famous phrase 'It is raining DNA outside', applied to downy seeds of the willow tree at the start of chapter 5 of *The Blind Watchmaker*, is poetic. But very little of that DNA leads anywhere; it's just another molecule to be broken down as the falling seeds rot. A few seeds survive to germinate; fewer still produce plants; and most of those die or are eaten before they grow into a willow tree and produce the next rainfall of seeds. DNA has to be in the right place (in sexual species, eggs or sperm) at the right time (fertilisation) before it propagates itself in any genetic sense. None of this stops genetics being a real science, and a very exciting and important one. So the fuzziness of definitions is not a good stick with which to beat the memetic dog, or indeed any dog that has anything going for it.

In his original discussion, almost as an aside, Dawkins suggested that religion is a meme, which goes something like 'If you wish to avoid the everlasting fires, you must believe *this*, and pass it on to your children'.* The popularity of religion is no doubt more complicated than that; nevertheless, there is the germ of an idea here, because that sentence does correspond rather closely to the central message of many – not all – religions. The theologian John Bowker was sufficiently disturbed by this suggestion that he wrote *Is God a Virus?* to shoot it down. The fact that he bothered shows that he saw it as an important (and from his viewpoint dangerous) question.

Blackmore recognises that a religion, or any ideology, is too complex to be propagated by a single meme, just as an organism is too complex to be propagated by a single gene. Dawkins recognised this, too, and came up with a concept that he called 'coadapted meme complexes'. These are systems of memes that replicate collectively. The meme 'If you wish to avoid the everlasting fires, you must believe *this*, and pass it on to your children' is too simple to get very far, but if it is allied to other memes like 'The way to avoid the everlasting fires

---

* The 'Shema' prayer, which orthodox Jews must say at least three times a day, includes 'And these words, which I command you this day, shall be upon your heart; and you shall teach them diligently to your children, and you shall talk of them when you sit in your house, when you walk upon the way, when you lie down and when you rise up.'

can be found in the Holy Book' and 'You must read the Holy Book or face eternal damnation', then the whole collection of memes forms a network that replicates far more effectively.

A complexity theorist would call such a collection of memes an 'autocatalytic set': each meme is catalysed, its replication is assisted, by some or all of the others. In 1995 Hans-Cees Speel coined the term 'memeplex'. Blackmore has a whole chapter on 'Religions as meme-plexes'. If this line of argument bothers you, hang on a minute. Are you saying that religion is *not* a collection of beliefs and instructions that can be passed very successfully from one person to another? That's what 'memeplex' means. Anyway, replace 'religion' by 'politi-cal party' if you want to – not the one you support, naturally. Those other idiots who advocate/despise (delete whichever is inapplicable) free market economics, state pensions, public ownership of industry, private ownership of public services … And bear in mind that while the secret of the spread of your own religion may be that it is The Truth, that can't possibly be the secret of the spread of all those other false religions in the world. Why the devil do sensible people believe *that* kind of rubbish?

Because it is a successful memeplex.

The evidence for memetic transmission of ideologies is extensive. For example, every one of the world's religions (barring ancient ones whose origins are lost in the mists of time) seems to have started with a very small group of believers and a charismatic leader. They are spe-cific to particular cultural backgrounds; the meme needs a fertile substrate on which to grow. Many cherished beliefs of Christianity, for example, seem absurd to anyone not brought up in the Christian tradition. Virgin birth? (Well, that one was actually an inspired mis-translation of the Hebrew for 'young woman', but no matter.) Restored the dead to life? Communion wine becomes blood? Communion wafers are the body of Christ – and you *eat* them? Really? To believers, of course, all this makes perfect sense, but to outsiders, uninfected by the meme, it's laughable.*

---

* Of course it ceases to be laughable if, despite its bizarre appearance, it happens to be *true*. And we've already agreed that all religions are true, for a given value of 'true'.

Blackmore points out that when it comes to a choice between doing good and spreading the meme, religious people tend to go for the meme. To most Catholics, and many other people, Mother Teresa was a saint (and she looks well set to become one in the fullness of time). Her work in the slums of Calcutta was selfless and altruistic. She did a lot of good, no question. But some Calcuttans feel that she diverted attention away from the real problems, and helped only those who accepted the teachings of her faith. For example, she was staunchly against birth control, the one practical thing that would have done the most good for the young women who needed her help. But the Catholic memeplex forbids birth control, and in a crunch, the meme wins. Blackmore sums up her analysis like this:

> These religious memes did not set out with an intention to succeed. They were just behaviours, ideas and stories that were copied from one person to another ... They were successful because they happened to come together into mutually supportive gangs that included all the right tricks to keep them safely stored in millions of brains, books and buildings, and repeatedly passed on to more.

In Shakespeare, memes become art. And now we move up another conceptual level. In drama, genes and memes cooperate to produce a temporary construct on a stage, for other extelligences to view. Shakespeare's plays give them pleasure, and change their minds. They, and works like them, redirect human culture by attacking our own mental elvishness.

The power of story. Don't leave home without it. And never, never, *never* underestimate it.

THIRTY-ONE

A WOMAN ON ſTAGE?

IT WAS THE SMELL of the theatre Rincewind remembered. People talked about 'the smell of the greasepaint, the roar of the crowd' but, he assumed, the word 'roar' must have been taken to mean the same as 'stink'.

He also wondered why this theatre was called The Globe. It was not even completely circular. But, he supposed, the new world might happen here …

He'd made a big concession for the occasion. He'd unstitched the few remaining sequins from the word 'WIZZARD' on his hat. Given its general lack of shape, and his robe's raggedness, it now made him look far more like one of the crowd, albeit a one that knew the meaning of the word 'soap'.

He worked his way back through the throng to the wizards, who had managed to get real seats.

'How is it going?' said Ridcully. 'Remember, lad, the show *must* go on!'

'Things are fine, as far as I can see,' whispered Rincewind. 'No sign of any elves at all. We did spot a fishmonger in the crowd, so the Librarian slugged him and hid him behind the theatre, just in case.'

'You know,' said the Chair of Indefinite Studies, who was leafing through the script, 'this chap would write much better plays if he didn't have to have actors in them. They seem to get in the way all the time.'

'I read the *Comedy of Errors* last night,' said the Dean. 'And I could see the error right there. There wasn't any comedy. Thank gods for directors.'

The wizards looked at the crowd. It wasn't as well behaved even as the ones back home; people were picnicking, small parties were being held, and there was a general sense that the audience looked upon the actual play as pleasant background noise to their personal social occasions.

'How will we know when it starts?' said the Lecturer in Recent Runes.

'Oh, trumpets get blown,' said Rincewind, 'and then generally two actors come on and tell one another what they already know.'

'No sign of the elves anywhere,' said the Dean, looking around with a hand over one eye. 'I don't like it. It's too quiet.'

'No, sir, no, sir,' said Rincewind. 'That's not the time not to like it. The time not to like it is when it's suddenly as noisy as all hell, sir.'

'Well, you get backstage with Stibbons and the Librarian, will you?' said Ridcully. 'And try not to look conspicuous. We mustn't take any chances.'

Rincewind worked his away around behind the stage, trying not to look conspicuous. But it was a first night, and there was an informality about the whole business that he'd never seen back home. People just seemed to wander around. Back home, there never seemed to be so much *pretence*; here, the actors played at being people and, down below, people played at being an audience. The overall effect was rather pleasing. The plays had a conspiratorial quality. Make it interesting enough, their audience was saying, and we'll believe anything. If you don't, we'll have a party with our friends right here and throw nuts at you.

Rincewind sat down on a pile of boxes offstage and watched as the play began. There were raised voices and the gentle, subtle sound of an expectant audience ready to tolerate quite a lot of plot exposition provided there was a joke or a murder at the end of it.

There was no sign of elves, no telltale shimmer in the air. The play wound on. Sometimes there was laughter, in which the deep boom of Ridcully was distinctly noticeable, especially, for some reason, when the clowns were on stage.

The stage elves met with approval, too. Peaseblossom, Cobweb, Moth and Mustardseed ... creatures of blossom and air. Only Puck seemed to Rincewind to be anything like the elves he knew, and even he seemed more of a prankster than anything else. Of course, the elves could be pranksters, too, especially if a footpath ran beside a really *dangerous* ravine. And the glamour they used ... well, here it was charming ...

... and there was the Queen, a few feet away. She didn't flash into existence, she emerged from the scenery. A group of lines and shadows that had always been there suddenly, without actually changing, became a figure.

She was wearing a black lace dress hung about with diamonds, so that she looked like walking night.

She turned to Rincewind, with a smile.

'Ah, potato man,' she said. 'We see your wizardly friends out there. But they won't be able to do anything. This show *will* go on, you know. Just as written.'

'... will go on ...' murmured Rincewind. He couldn't move. She'd hit him with her full force. In desperation, he tried to fill his mind with potatoes.

'We *know* you told him a garbled version,' said the Queen, walking around his quivering body. 'And a lot of nonsense it was. So I appeared to him in his room and put the whole thing in his mind. So simple.'

Roast potatoes, thought Rincewind. Sort of gold with brown edges, and maybe almost black here and there so they're nice and crunchy ...

'Can't you hear the applause?' said the Queen. 'They like us. They actually *like* us. We'll be in their paintings and stories from now on. You'll *never* get us out of there ...'

Chips, thought Rincewind, straight from the deep fryer, with little bubbles of fat still spitting and popping ... but he couldn't stop his treacherous head from nodding.

The Queen looked puzzled.

'Don't you think about *anything* but potatoes?' she said.

Butter, thought Rincewind, chopped chives, melted cheese, salt ...

But he couldn't stop the thought. It opened up inside his head, pushing away all potato-shaped fantasies. *All we have to do is nothing, and we've won!*

'What?' said the Queen.

Mash! Huge mounds of mash! Creamed mash!

'You're trying to hide something, wizard!' said the Queen, a few inches from his face. 'What is it?'

Potato cakes, fried potato skins, potato croquettes …

… no, not potato croquettes, no one ever did them properly … and it was too late, the Queen was reading him like a book.

'So …' she said. 'You think only mysteries last? Knowledge in *unbelief?* Seeing is *disbelieving?*

There was a creaking above them.

'The play's not over, wizard,' said the Queen. 'But it's going to stop right *now*.'

At this point, the Librarian dropped on her head.

Winkin the glove stitcher and Coster the apple seller discussed the play on the way home.

'The bit with the queen and the man with the *aʃʃeʃ earʃ waʃ* good,' said Winkin.

'Aye, it waʃ.'

'And the wall bit, too. When the man ʃaid "he iʃ no creʃcent, and hiʃ hornʃ are inviʃible within the circumference", I nearly widdled my breecheʃ. I like a good joke, me.'

'Aye.'

'But I didn't underʃtand why all thoʃe people in the fur and featherʃ and ʃtuff were chaʃed acroʃs the ʃtage by the man in the hairy red coʃtume, and why the fat men in the expenʃive ʃeats all got up and on to the ʃtage and why the idiot in the red dreʃs waʃ running around ʃcreaming about potatoʃ, whatever *they* are. While Puck waʃ ʃpeaking at the end I definitely thought I could hear a fight going on.'

'Experimental theatre,' said Winkin.

'Good dialogue,' said Coster.

'And you've got to hand it to thoʃe actorʃ, the way they kept going,' said Winkin.

'Yeah, and I could have ʃworn there was another Quene up on stage,' said Coster, 'and ʃhe looked like a woman. You know, the one who was trying to ʃtrangle that man babbling about potatoeʃ.'

'A woman on ſtage? Don't be daft,' said Winkin. 'Good play, though.'

'Yeah. I think they could cut out the chaſe ſequence, though,' said Coster. 'And frankly I don't think you could get a girdle that big.'

'Yes, it would be dreadful if ſpecial effectſ took over,' said Winkin.

Wizards, like many large men, can be quite light on their feet. Rincewind was impressed. By the sound of it, they were right behind him as he sped along the path by the river.

'Best not to wait for a curtain call, I thought,' Ridcully panted.

'Did you see me … wallop the Queen with a horseshoe?' wheezed the Dean.

'Yes … pity it was an actor,' said Ridcully. 'The other one was the elf. Still, not a *complete* waste of a horseshoe.'

'But we certainly showed them, eh?' said the Dean.

'The history is completed,' said the voice of Hex, from Ponder's bouncing pocket. 'Elves will be viewed as fairies and such they will become. Over the course of several centuries belief in them will dwindle as they are moved into the realm of art and literature, which is where the remnant of them will subsequently exist. They will become a subject suitable for the amusement of children. Their influence will be severely curtailed but will never die away completely.'

'*Never*?' panted Ponder, who was getting winded.

'There will always be some influence. Minds on this world are extremely susceptible.'

'Yes, but we've pushed imagination to the next stage,' puffed Ponder. 'People can imagine that the things they imagine are imaginary. Elves are little fairies. Monsters get pushed off the map. You can't fear the unseen when you can see it.'

'There will be new kinds of monsters,' said Hex, from Ponder's pocket. 'Humans are very inventive in that respect.'

'Heads … on … spikes,' said Rincewind, who liked to save his breath for running.

'Many heads,' said Hex.

'There's always heads on spikes somewhere,' said Ridcully.

'The Shell Midden People didn't have heads on spikes,' said Rincewind.

'Yes, but they didn't even have spikes,' said Ridcully.

'You know,' wheezed Ponder, 'we *could* have just told Hex to move us directly to the opening into L-space ...

They landed on the wooden floor, still running.

'Can we teach him to do that on Discworld?' said Rincewind, after they'd picked themselves up from the heap by the wall.

'No! Otherwise what use would you be?' said Ridcully. 'Come on, let's go ...'

Ponder hesitated by the L-space portal. It was filled with dull, greyish light, and a distant view of mountains and plains of books.

'There's still elves here,' he said. 'They're persistent. They might find some way to—'

'Will you come on?' snapped Ridcully. 'We can't fight *every* battle.'

'Something could still go wrong, though.'

'Whose fault will that be now? No, come *on*!'

Ponder looked around, gave a little shrug, and stepped into the hole.

After a moment a hairy red arm came through and pulled more books through the hole, piling them up until it was a wall of books.

Brilliant light, so strong that it lanced out between the pages, flashed for a while somewhere in the heap.

Then it went dark. After a moment, a book slipped out of the pile, and it collapsed, the books tumbling to the floor, and there was nothing left but a bare wall.

And, of course, a banana.

THIRTY TWO

MAY CONTAIN NUTS

 WE ARE THE STORYTELLING APE, and we are incredibly good at it.

As soon as we are old enough to want to understand what is happening around us, we begin to live in a world of stories. We *think* in narrative. We do it so automatically that we don't think we do it. And we have told ourselves stories vast enough to live in.

In the sky above us, patterns older than our planet and unimaginably far away have been fashioned in gods and monsters. But there are bigger stories down below. We live in a network of stories that range from 'how we got here' to 'natural justice' to 'real life'.

Ah, yes … 'real life'. Death, who acts as a kind of Greek chorus in the Discworld books, is impressed by some aspects of humanity. One is that we have evolved to tell ourselves interesting and useful little lies about monsters and gods and tooth fairies, as a kind of prelude to creating really big lies, like 'Truth' and Justice'.

There is no justice. As Death remarks in *Hogfather*, you could grind the universe into powder and not find one atom of justice. We created it, and while we acknowledge this fact, nevertheless there is a sense in which we feel it's 'out there', big and white and shining. It's another story.

Because we rely so much on them, we love stories. We require them on a daily basis. So a huge service industry has grown up over several thousand years.

The basic narrative forms of drama – the archetypal stories – can all

340

be found in the works of the ancient Greek playwrights: Aeschylus, Aristophanes, Euripides, Sophocles ... Most of the dramatic tricks go back to ancient Greece, especially Athens. No doubt they are older than that, for no tradition starts in fully developed form. The 'chorus', a gaggle of bit-players who form a backdrop for the main action and in various ways reinforce it and comment on it, is of Greek or earlier origin. So is the main division of the form of a play, though not necessarily its substance, into comedy and tragedy. So, possibly, is the invention of the huge stuffed joke willy, always good for a laugh from the cheap seats.

The Greek concept of tragedy was an extreme form of narrative imperative: the nature of the impending disaster had to be evident to the audience and to virtually all of the players; but it also had to be evident that it was going to happen anyway, despite that. You were Doomed, as you should be – but we'll watch anyway, to see how *interestingly* you'll be Doomed. And if it sounds silly to watch a drama when you know the ending in advance, consider this: how likely is it, when you settle down to watch the next James Bond movie, that he *won't* defuse the bomb? In fact you'll be watching a narrative as rigid as any Greek drama, but you'll watch anyway to see how the trick is done this time.

In our story, Hex is the chorus. In form, our tale is comedy; in substance, it is closer to tragedy. The elves are a Discworld reification of human cruelty and wickedness, they are evil incarnate because – traditionally – they have no souls. Yet in their various aspects they fascinate us, as do vampires and monsters and werewolves. It'd be a terrible event if the last jungle yields up its tiger, and so it would be, too, when the last forest yields up its werewolf (yes, all right, technically they don't exist, but we hope you know what we mean: it'd be a bad day for humanity when we stop telling stories).

We've piled on to elves and yetis and all the other supernatural aspects of ourselves; we're happier to say that monsters are out there in the deep dark forest than locked in here with us. Yet we need them, in a way we find hard to articulate; the witch Granny Weatherwax tried to summarise it in *Carpe Jugulum*, when she said 'We need vampires, if only to remind us what garlic is for'. G.K. Chesterton did rather better when, in an article defending fairy stories, he disputed the suggestion that stories tell children that there are monsters. Children already *know* there are monsters, he said. Fairy stories tell them that monsters can be killed.

We need our stories to understand the universe, and sometimes we forget that they're only stories. There is a proverb about the finger and the moon; when a wise man points at the Moon, the fool looks at the finger. We call ourselves *Homo sapiens*, possibly out of a hope that this may be true, but the storytelling ape has a tendency to confuse moons and fingers.

When your god is an ineffable essence that exists outside of space and time, with unimaginable knowledge and indescribable powers, a god of boundless sky and high places, belief slips easily into the mind.

But the ape isn't happy with that. The ape gets bored with things it can't see. The ape wants pictures. And it gets them, and then a god of endless space becomes an old man with a beard sitting in the clouds. Great art takes place in the god's honour, and every pious brush gently kills what it paints. The wise man says 'But this is just a metaphor!', and the ape says 'Yeah, but those tiny wings couldn't lift a cherub that fat!' And then not so wise men fill the pantheon of heaven with hierarchies of angels and set the plagues of man on horseback and write down the dimensions of Heaven in which to imprison the lord of infinite space.* The stories begin to choke the system ...

Seeing is *not believing*.

Rincewind knows this, which is why he encourages Shakespeare to make elves real. Because once you're called Mustardseed, it's downhill all the way ...

The elves cannot understand Rincewind's ploy. Not until his thoughts give it away to the Queen of the Elves, and the salvation of the world rests upon 300 pounds of plummeting orangutan. Nevertheless, the plan worked very well. This is Oberon, near the end of the play:

---

* Revelation xxi.16 gives it as 12,000 furlongs in length, breadth and height, or a cube 1,500 miles on a side. Noticeably smaller than the Moon.

Through the house give glimmering light,
By the dead and drowsy fire;
Every elf and fairy sprite
Hop as light as bird from brier;
And this ditty, after me,
Sing and dance it trippingly.

There's no hope for them. Next stop, nursery wallpaper. Whereas witches, now:

Scale of dragon, tooth of wolf,
Witch's mummy, maw and gulf
Of the ravin'd salt-sea shark,
Root of hemlock digg'd I'th' dark,
Liver of blaspheming Jew,
Gall of goat, and slips of yew
Silver'd in the moon's eclipse,
Nose of Turk, and Tartar's lips,
Finger of birth-strangled babe
Ditch-delivered by a drab—
Make the gruel thick and slab;
Add thereto a tiger's chaudron,
For th' ingredience of our cauldron.

No contest. What's a chaudron? Entrails. *Definitely* no contest. The witches appear on stage in *Macbeth* only three times, but they steal the show. They probably got fan mail. The fairies are present for a large part of *A Midsummer Night's Dream*, but it is Bottom that steals the show and only Puck has a glimmer of the old evil. They've been parcelled, stamped and sent on their way to Tinkly Wood.

To be sure, Shakespeare's Oberon is not all sweetness and light. He uses the juice of a herb, the flower known as Love-in-idleness, to enchant Titania, Queen of the Fairies, because she has gained possession of a changeling child, and he wants it. He makes her fall in love with Bottom, who at that point in the story is an ass. And he is appeased, and she is entirely happy with the turn of events, when she gives him the child. But that's low-level, sanitised nastiness, a fretful squabble, not a war.

The allure of the unknown fades into the tawdry reality of a specific representation, once you see it dripping sequins. Abraham's God of Extelligence was far more compelling than a few golden (probably just gold leaf) idols. But when the Renaissance artists started to paint God as a bearded man in the clouds, they opened the way to doubt. The image just wasn't impressive enough. The pictures on radio are always so much better than those on TV.

For the last few hundred years, humanity has been killing its myths. Faith and superstition have been giving way, slowly and against considerable resistance, to the critical assessment of evidence. They may, perhaps, be enjoying a bit of a revival: many rational thinkers have bemoaned the slide into cults and the wierd offshoots of New Ageism ... But those are all very subdued versions of the old myths, the old beliefs; their teeth have been drawn.

Science alone is not The Answer. Science too has its myths. We have shown you some of them, or at least what we believe to be some of them. The misuse of anthropic reasoning is a clear example, as in the case of the carbon resonance, but argued with no thought for the fudge-factor of the red giant.

The ideal of the scientific method is often not realised. Its usual statement is an oversimplification in any case, but the basic worldview captures the essence. Think critically about what you are told. Do not accept the word of authority unthinkingly. Science is not a belief system: no belief system instructs you to question the system itself. Science does. (There are many scientists, however, who treat it as a belief system. Be wary of them.)

The most dangerous myths and ideologies, today, are the ones that have not yet been destroyed by the rising ape. They still strut their stuff on the world's stage, causing grief and havoc – and the tragedy is that it's all to no purpose. Most of it doesn't *matter*. Issues like abortion do matter, to some extent; even 'pro-choice' adherents would prefer that the choice should not be necessary. Issues like short skirts or lengths of beards do *not* matter, and it's foolish and dangerous to make a big fuss about them on a planet that is bursting at the seams with an excess of people. To do so is to promote the memeplex above the good of humanity. It is the action of a barbarian mind, a mind sufficiently removed from reality that the consequences of its resident memeplex do not affect it directly. It is not the actions of the naïve young men who carry the suicide bomb, or fly the airliner into a skyscraper, that are the root of the problem; it is the actions of the evil old men who lead them to behave like that, all for the sake of a few memes.

The key memes are not religious, in this case, we suspect, even though religion is often blamed: that's mostly a smokescreen. Those old men are motivated by political memes, and the religious meme-plex is merely another of their weapons. But they are also trapped in their own stories, and this is high tragedy. Granny Weatherwax would never make that mistake.

The elves are still with us, in our heads. Shakespeare's humanity, and the critical faculties encouraged by science, are two of our weapons against them. And fight them we must.

And to achieve that, we need to invent the right stories. The ones we've got have brought us a long way. Plenty of creatures are intelligent, but only one tells stories. That's us, *Pan narrans*.

And what about *Homo sapiens*? Yes, we think that would be a very good idea …

INDEX